AF564907

ENCYCLOPAEDIA OF ATOMIC PHYSICS

Encyclopaedia of
Atomic Physics

Volume 1

Shehzad Ahmad

ANMOL PUBLICATIONS PVT. LTD.
NEW DELHI - 110 002 (INDIA)

ANMOL PUBLICATIONS PVT. LTD.
H.O.: 4374/4B, Ansari Road, Darya Ganj,
New Delhi-110 002 (India)
Ph.: 23278000, 23261597
B.O.: No. 1015, Ist Main Road, BSK IIIrd Stage
IIIrd Phase, IIIrd Block
Bangalore - 560 085 (India)
Visit us at: www.anmolpublications.com

Encyclopaedia of Atomic Physics

First Published, 2007

ISBN 81-261-3072-5 (Set)

PRINTED IN INDIA

Printed at Mehra Offset Press, Delhi.

Contents

Preface

Atom was considered as the smallest and indivisible particle of matter till some decades back. But scientists have further succeeded in dividing it too. An atom consists of three parts, electron, proton and neutron. Of them, electron is the most important one. It always remains in motion yielding energy. It is this electron, which has given birth to a new world called electronic world, which is millions of years faster than the non-electronic one.

Atomic physics relates to the physical properties and functions of atoms; their relation to other particles; effects of atoms on the mechanism of natural phenomena and vice-versa. After attempts of years, scientists succeeded in breaking the nucleus of an atom which released huge energy. It consists of two subatomic particles, called lower quark and upper quark. They are the building blocks of proton and neutron.

Now, scientists are endeavouring to break even the quark. So far, they have not succeeded but their efforts are on. They are hopeful they would succeed in their endeavour one day or the other.

Encyclopaedia of Atomic Physics is an important piece of work consisting of three volumes. It attempts at highlighting the phenomena which involve atomic physics at length. Almost every aspect has been taken note of, while exploring the vast world of atomic physics.

Hopefully, it would prove to be an asset for the scholars, students, researchers, academics and the general readers, alike.

We invite and appreciate the comments and suggestions sent by the readers.

— **Editor**

1

Mode of Atom

Dividing Particle

In 1906, Rutherford observed that when a sharp beam of α-particles falls upon a photographic plate in vacuum, a sharp image is obtained. If, however, a thin foil of metal is placed in the path of the beam, the image becomes diffuse. This is due to the scattering of α-particles by the atoms of the foil.

In 1909, Geiger and Marsden made a series of measurements on the scattering of α-particles. They placed a thin metal gold foil in the path of a ray of α-particles from a radioactive substance (Fig. below). A fluorescent screen S, backed by a microscope M, was placed behind the foil. The particles, while passing through the foil deflected (scattered) through a wide range of angles.

The number of particles scattered in various scattered in various directions were obtained by counting the scintillations they made on the screen S. The screen was always maintained perpendicular to the incident ray. It was found that *although most of the particles scattered through angles of the order of 1° or less, out a small number, say about 1 in every 10,000, scattered through 90° or even 180°*. This *large-angle* scattering of α-particles could not be at all explained from the-then prevalent ideas about atom.

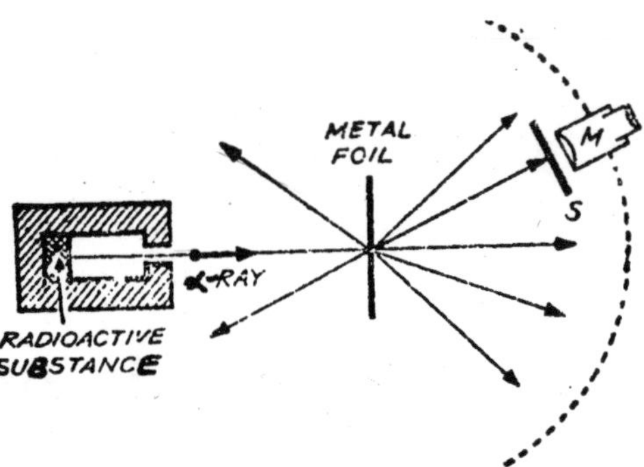

In 1911, Rutherford explained, successfully the large-angle scattering by proposing a structure of the atom. He made the following assumptions :

(i) The positive charge of the atom is concentrated in an extremely small volume, known as nucleus', in the atom negative charges (electrons) revolve around the nucleus in planetary orbits. Thus most of the atom consists of empty space. This is why a great majority of α-particles pass the metal foil almost undeviated.

(ii) The whole mass of the atom (excluding the electrons) is contained in the nucleus along with the positive charge.

(iii) The scattering of α-particles by atoms takes place due to an encounter between the a-particle and the *nucleus* of the atom. The electrons outside the nucleus, owing to their little mass, cannot appreciably deflect the far more massive α-particles.

(iv) Both the α-particle and the nucleus are small enough to be considered as *point* mass and *point* (positive) charge; and the electrostatic repulsive force between them is governed by Coulomb's inverse-square law,

(v) The nucleus of the atom (of the metal foil used) is so massive compared with the α-particle that it remains at rest during the encounter.

When an α-particle strikes the metal foil, it can penetrate the outer electron-cloud of an atom and approach closely the nucleus. It then moves under the action of a coulombian force of repulsion, and its path is a *hyperbola* with the nucleus as the external focus.

Theory of Rutherford

The Rutherford's assumptions regarding α-scattering have been given in the last question.

Let us consider an α-particle moving in a direction PO (Fig.), where *P* is far from a stationary nucleus *N*. As the particle approaches the nucleus, it begins to be repelled more and more and assumes the hyperbolic path *PAP'*. *OP* and *OP'* make equal angles, *6*, with the line *NA* where *A* is the apex of the hyperbola. Let *p* be the perpendicular distance from the

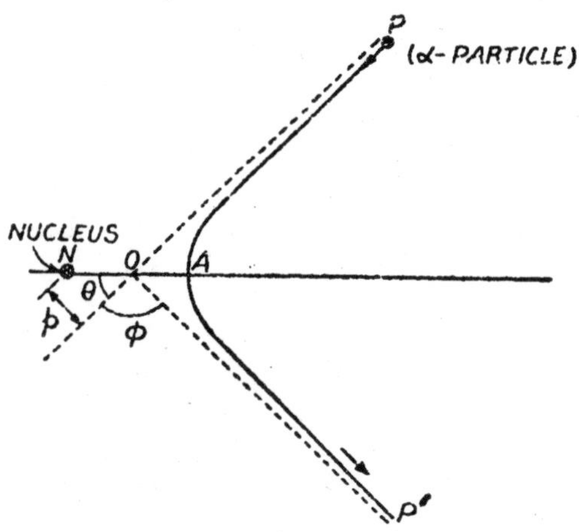

nucleus *N* to the original direction of the α-particle. *v* is called the 'impact parameter'. It is the minimum distance to which the α-particle would approach the nucleus if there were no force between them. The angle between the direction of approach and the direction of recede of the α-particle· is the 'scattering angle' ϕ. Let us first find a relation between p and ϕ.

Let m_α. be the mass of the α-particle, v its velocity at P, and v' the velocity at A. By conservation law, the initial angular momentum* of the α-particle about the nucleus will be equal to its angular momentum when at A. Therefore

$$m_x \, v \, P = m_\alpha \, v' \, a,$$

where $$a = NA.$$

Thus

$$v = \frac{v\,p}{a} \qquad \text{... } (i)$$

We know that the (positive) charge on the nucleus is Ze, where Z is the atomic number, and that on the α-particle is $2e$, where e is the electronic charge. Therefore, the electrostatic potential energy of the α-particle when at A will be $\frac{1}{4\pi_0^\epsilon}$ $\frac{(Ze)(2e)}{a}$. Now, from the conservation of energy, the initial kinetic energy of the α-particle is equal to the sum of the kinetic and potential energies at A, so that

$$\frac{1}{2} m_\alpha \, v^2 = \frac{1}{2} m_\alpha \, v^2 + \frac{1}{4\pi_0^\epsilon} \frac{2Ze^2}{a}$$

or $$v'^2 = v^2 - \frac{1}{4\pi_0^\epsilon} \frac{4Ze^2}{m_\alpha \, a}$$

or $$v'^2 = v^2 \left(1 - \frac{1}{4\pi_0^\epsilon} \frac{4Ze^2}{m_\alpha \, c^2 \, a}\right)$$

Let us introduce $E_k \, (= \frac{1}{2} m_\alpha \, v^2)$, *the initial* kinetic energy of the α-particle, instead of its mass m_z and velocity v. Then we have

$$v'^2 = v^2 \left(1 - \frac{1}{4\pi_0^{\epsilon}} \frac{4Ze^2}{E_k\, a}\right)$$

or
$$v'^2 = v^2 \left(1 - \frac{b}{a}\right),$$

where
$$b = \left(\frac{1}{4\pi_0^{\epsilon}} \frac{2Ze^2}{E_k}\right).$$

Now, putting the value of v' from eq. (*i*), we get

$$\frac{v^2 p^2}{a^2} = v^2 \left(1 - \frac{b}{a}\right),$$

or
$$p^2 = a^2 - ba.$$

N is the external focus and A the apex of the hyperbola. It can be shown from the coordinate geometry that

$$a = NA = p \cot (\theta/2),$$

where θ is the angle shown. Therefore the above expression becomes

$$p^2 = p^2 \cot^2 (\theta/2) - bp \cot (\theta/2).$$

$$\therefore \qquad b = p\, \frac{\cot^2 (\theta/2) - 1}{\cot (\theta/2)}$$

$$= p \{\cot (\theta/2) - \tan (\theta/2)\}$$

$$= p \left\{\frac{\cos (\theta/2)}{\sin (\theta/2)} - \frac{\sin (\theta/2)}{\cos (\theta/2)}\right\}$$

$$= p \left\{\frac{\cos^2 (\theta/2) - \sin^2 (\theta/2)}{\sin (\theta/2) \cos (\theta/2)}\right\}$$

$$= 2p \left\{\frac{\cos \theta}{\sin \theta}\right\} = 2p \cot \theta.$$

$$\therefore \qquad \cot \theta = \frac{b}{2p}. \qquad \text{...}(ii)$$

From the fig., the scattering angle ϕ is equal to $\pi—2\theta$, so that

$$\frac{\phi}{2} = \frac{\pi}{2} - \theta$$

or $$\cot\frac{\phi}{2} = \cot\left(\frac{\pi}{2} - \theta\right) = \tan\theta.$$

But tan $$\theta = \frac{2p}{b} \text{ from eq. } (ii).$$

$$\therefore \qquad \cot\frac{\phi}{2} = \frac{2p}{b}. \qquad \ldots (iii)$$

This shows that the scattering angle ϕ decreases *rapidly* as the impact parameter p increases. It means that *only an α-particle passing extremely close to the nucleus suffers large-angle scattering.*

Eq. (*iii*) cannot be verified experimentally because we cannot measure the impact parameter p for an observed angle ϕ. We therefore calculate the number of α-particles scattered through an angle ϕ. We see from eq. (iii) that all α-particles, such as 1, 2, 3 in the below–given figure , which approach the nucleus with an impact parameter less than p will be scattered through an angle greater than ϕ. This means that an α-particle that is initially directed anywhere within the area πp^2 around the nucleus will be scattered through an angle greater than ϕ. Let t be the thickness of the scattering foil and n the number of atoms per unit volume. Then, it can be shown from the theory of probability that the fraction f of incident α-particles scattered through angle greater than ϕ is

$$f = \pi p^2 nt.$$

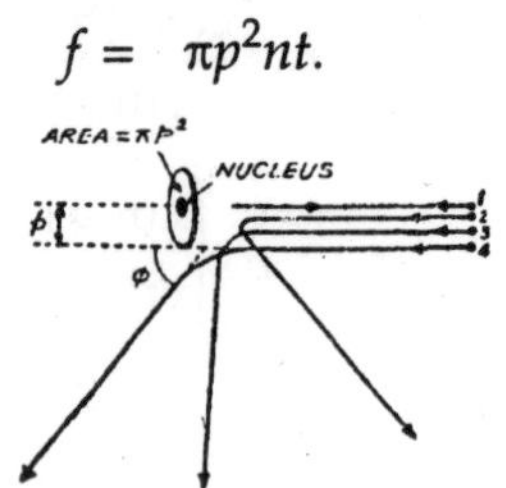

We assume that the foil is sufficiently thin so that an α-particle receives the entire deflection from a single encounter with a nucleus. Then we can substitute the value of p $\left(= \frac{b}{2} \cot \frac{\phi}{2}\right)$ from eq. (*iii*). Thus

$$f = \frac{1}{4} \pi b^2 nt \cot^2 \frac{\phi}{2}.$$

Therefore, the fraction of incident α-particles scattered between ϕ and $\phi + d\phi$ is

$$df = \frac{1}{4} \pi b^2 nt \cot \left(\frac{\phi}{2}\right) \text{cosec}2 \left(\frac{\phi}{2}\right) d\phi. \quad \text{... } (iv)$$

The minus sign indicates that f decreases with increasing ϕ.

In the experiment, a fluorescent screen was placed at a distance r from the foil, and the scattered α-particles were counted by means of the scintillations they produced on the screen. Those particles which scattered between ϕ and $\phi + d\phi$ reach a spherical zone of radius $r \sin \phi$ and width $r\, d\phi$ (Fig.). Thus the area dS of the screen struck by these particles is

$$DS = (2\pi\, r \sin \phi)\, (r\, d\phi)$$

$$= 2\pi\, r^2 \sin \phi\, d\phi$$

$$= 4\pi\, r^2 \sin (\phi/2) \cos (\phi/2)\, d\phi. \quad \text{... } (v)$$

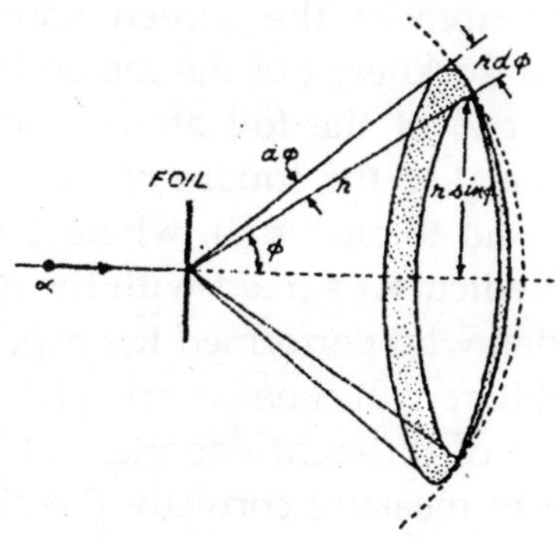

If N_0 be the total number of α-particles striking the foil during the experiment, the number scattered between ϕ and $\phi + d\phi$ is $N_0\, df$. Therefore, the number $N(\phi)$ counted per unit area on the screen placed at angle ϕ, which is the quantity actually measured, is given by

$$N(\phi) = \frac{N_0 |df|}{dS}.$$

Putting the values of df and dS from eq. (iv) and (v), we get

$$N(\phi) = \frac{N_0 (1/4)\pi b^2\, nt \cot(\phi/2) \operatorname{cosec}^2(\phi/2)\, d\phi}{4\pi r^2 \sin(\phi/2)\cos(\phi/2)\, d\phi}$$

$$= \frac{N_0 b^2\, nt}{16 r^2} \frac{1}{\sin^4(\phi/2)}$$

Substituting the value of $b \left(= \frac{1}{4\pi\epsilon_0}\frac{2Ze^2}{E_k}\right)$, we obtain

$$\boxed{N(\phi) = \frac{N_0\, ntZ^2 e^4}{4r^2 (4\pi\epsilon_0)^2 E_k^2} \frac{1}{\sin^4(\phi/2)}.} \qquad ...(vi)$$

This is Rutherford's scattering formula for the number of α-particles scattered at a particular angle (ϕ).

Significance of Rutherford's Formula

The formula shows that the number of scattered α particles arriving per unit area at the screen should be directly proportional to the thickness t of the foil and to the square of the atomic number Z of the foil atoms ; and it should be inversely proportional to the square of the kinetic energy E_k of the α-particles and to $\sin^4(\phi/2)$, where ϕ is the scattering angle. All these predictions agreed with the measurements of Geiger and Marsden who performed the experiment by using scattering foils of different thicknesses and of different materials, and the α-particles of different energies. In 1920, Chadwick used the formula to measure correctly the charge Ze on the nucleus for a number of elements, thus giving a further verification of the formula. Hence we conclude that Rutherford's assumption of the existence of a tiny, massive, positive nucleus in the atom is correct. Rutherford is therefore credited with the "discovery" of the nucleus.

Angle of Scattering

The angle of scattering ϕ of an α-particle is related to the impact parameter p by

$$\cot \frac{\phi}{2} = \frac{2p}{b},$$

where $b = \frac{1}{4\pi_0^{\epsilon}} \frac{2Ze^2}{E_k}$, E_k being the initial kinetic energy of the particle.

Now, b is same for the two α-particles, because E_k is same. Larger the scattering angle ϕ, smaller will be the impact parameter p. Clearly, the particle scattered through a *larger* angle comes closer to the nucleus.

It is found that the number of α-particles scattered through very small angles is somewhat different from the number predicted by Rutherford's formula. The reason for the same is as follows.

Rutherford's formula fails to agree with the experimental data at very small scattering angles. This is because small ϕ implies a large impact parameter, in which case the full nuclear charge of the target atom in partially screened by its electrons.

Distance of Closest Approach : When α-particles falls upon a thin metal foil, the particles are scattered in various directions by the *nuclei* of the atoms of the foil. The angle of scattering of a particular particle depends upon the particle's impact parameter (the perpendicular distance from the nucleus to the initial direction of motion of the particle). Rutherford showed that if the nucleus be assumed to be a *point mass* and a *point charge* and the electrostatic force between the nucleus and the α-particle be assumed to be governed by Coulomb's inverse-square law, then the angle of scattering is related to the impact parameter p by the equation

$$\boxed{\cot\left(\frac{\phi}{2}\right) = \frac{2p}{b},}$$

where $b\left(=\frac{1}{4\pi\overset{\epsilon}{0}}\frac{2Ze^2}{E_k}\right)$ depends upon the charge of the given nucleus and the kinetic energy of the smaller the incident α particle. This smaller the impact parameter p, greater the scattering angle . When $p = 0$, then ϕ assumes the maximum value of 180°. This means that when the α-particle moves directly (head-on) toward the nucleus, it is scattered through 180°, *i.e., reflected back* along its initial path (Fig.). Clearly, it is under this condition that the particle will have its closest approach to the nucleus.

Let r_0 be the distance of closest approach of the α-particle to the nucleus. The (positive) charge on the nucleus is Ze, and that on the α-particle is $2e$, where e is the electronic charge. Therefore, the electrostatic potential energy of the particle at the instant of closest approach is $\frac{1}{4\pi\overset{\epsilon}{0}}\frac{(Ze)(2e)}{r_0} = \frac{1}{4\pi\overset{\epsilon}{0}}\frac{2Ze^2}{r_0}$. At this instant the α-particle is momentarily at rest, and the initial kinetic energy E_k is entirely converted into electrostatic potential energy. Hence at this instant

$$E_k = \frac{1}{4\pi\overset{\epsilon}{0}}\frac{2Ze^2}{r_0}$$

$$\boxed{\therefore\ r_0 = \frac{1}{4\pi\overset{\epsilon}{0}}\frac{2Ze^2}{E_K}.}$$

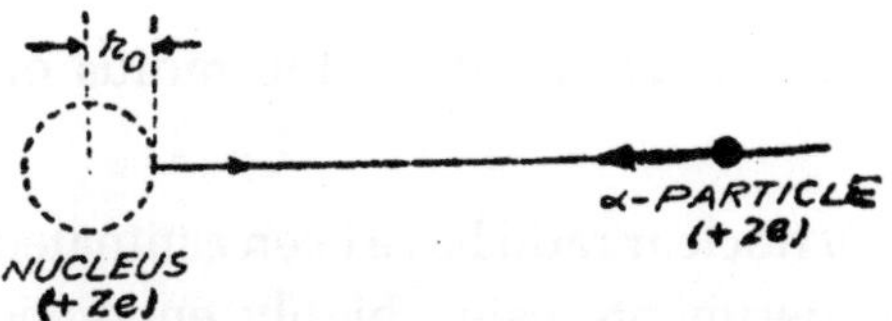

This is the expression for the distance of closest approach r_0 of the α-particle. It shows that for a given nucleus, r_0 depends upon the initial kinetic energy E_k of the α-particle.

When the kinetic energy E_k exceeds a certain value, the distance of closest approach r_0 becomes so small that the nucleus no longer appears as a *point* charge to the α-particle. Then the Coulomb's inverse-square law and hence the Rutherford formula breaks down. Besides this, some short-range *attractive* nuclear forces also begin to operate. Hence under this condition the incident particle is not reflected back by the nucleus; instead the particle penetrates the nucleus. Thus, *there is an upper limit for the energy of the incident particle beyond which the particle cannot be reflected back by a given nucleus.*

Estimation of Nuclear Size—Upper Limit of Nuclear Radius : The Rutherford scattering formula holds so long the nucleus appears as *a point* to the scattering α-particle *i.e.* so long the size of the nucleus is insignificant compared with the closest distance r_0 to which the α-particle approaches the nucleus. The more energetic the incident a-particle; the smaller is r_0. Hence the distance of the closest approach, r_0, of a particle which has a maximum energy (upto which the Rutherford formula agrees with experiment) is the upper limit for the nuclear radius.

The experimental observations on α-particle scattering are found to agree with formula upto a maximum energy of 7-8 MeV. For α-particles of this energy the distance of closest approach is about 3×10^{-14} meter for a gold nucleus and 2×10^{-14} meter for a silver nucleus. Hence we conclude that the radius of the gold nucleus is less than 3×10^{-14} meter, and that of the silver nucleus is less than 2×10^{-14} meter. In general, the nucleus is well under 1/10,000 the radius of the atom as a whole.

The *actual* nuclear radii have been estimated by performing scattering experiments using highly energetic electrons and neutrons and studying the deviation of the results from the Rutherford formula. It has been found that the nuclear radius R is given by

$$R = R_0 A^{1/3}$$

where A is the mass number of the nucleus and

$$R_0 \approx 1.3 \times 10^{-15} \text{ meter.}$$

Formation of Atom

Matter, despite its appearance of being continuous, possesses a definite particle structure of a microscopic level. The ultimate particles of matter are known as 'atoms'. Today we know that an atom has its own ultimate particles, electrons, protons, and neutrons.

The first information about the structure of atom came from the discovery of the electron in 1897, when it became known that all atoms contain the negatively-charged electrons. Since matter is electrically-neutral, it was inferred that atoms must also contain positively-charged material. Furthermore, an electron was found thousands of times lighter than the entire atom. This suggested that the positive charge of the atom carries nearly the entire mass. The radius of the atom was estimated to be of the order of 10^{-10} meter from kinetic theory considerations. The distribution of electron and positive charge in the atom was not known however.

Thomson's Model of Atom: Thomson, in 1904, suggested that an atom is a *uniform* sphere of positively-charged material, in which electrons are embedded in number such as to balance the positive charge (Fig.). The model was however, soon discarded as it was in complete disagreement with experiments on the *large-angle* scattering of α-particles by matter.

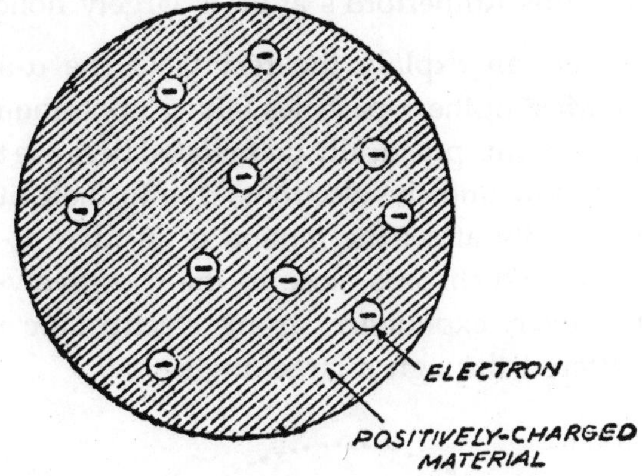

Rutherford's Nuclear Model of Atom: In 1909, Geiger and Marsden made a series of experiments on the scattering of α-particles. They found that α-particles falling on a thin metal foil scattered in various directions. Although most of the particles scattered through angles of the order of 1° or less, but a small number scattered through 90° or even 180°. This led Rutherford to conclude that while most of the (positively-charged) α-particles, while traversing matter, experienced little force, but a few of them experienced *strong* repulsive force as could be exerted by extremely dense concentrations of positive charge. Hence the positive charge of the atom could not be uniformly distributed as pictured by Thomson.

Rutherford suggested that in an atom the entire positive charge and nearly all of its mass are concentrated at the centre of the atom in a small volume, known as the 'nucleus' of the atom (Fig.). The electrons revolve around the nucleus in planetary orbits at distances large compared with the size of the nucleus. The orbital motion was assumed because without it the electrons would fall into the nucleus under the electrostatic attraction and the atom would collapse.

The radius of the nucleus was estimated to be of the order of 10^{-15} meter which is only 1/10,000 of the radius of the atom as a whole. Thus Rutherford's atom is largely hollow.

The model can explain the results of the α-scattering experiment. Most of the α-particles falling upon the metal foil go through the atoms practically undeviated because the atoms are largely hollow and the electrons, due to their little mass, do not appreciably affect the motion of incident α-particles. Those particles which pass close to the (positively-charged) nucleus, however, experience strong electrostatic repulsion and are scattered through a large angle.

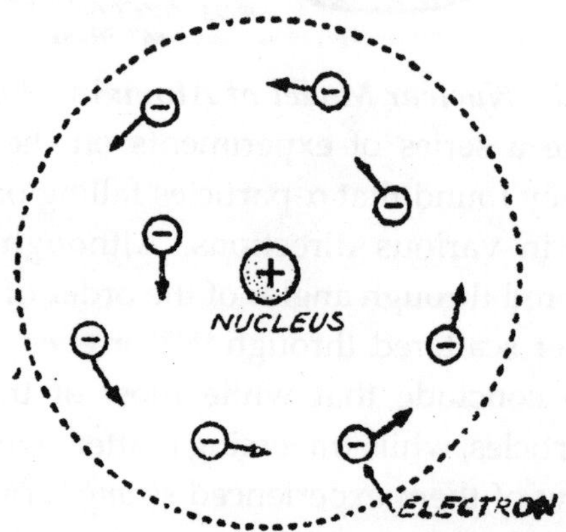

The Rutherford's model was supported by the periodic table of elements also. The α-scattering experiments enabled direct determination of the nuclear charge of the atom involved. It was found that all atoms of any one element had the same unique nuclear charge, and this charge increased regularly from element to element in the periodic table. The nuclear charge for an element was always found to be +*Ze*, where Z determined the position of the element in the periodic table. Z is now known as the 'atomic number' of the element, and gives the number of unit positive charges in a nucleus of that element.

Difficulties in Rutherford's Model (Failure of Classical Physics): The Rutherford's model, in spite of strong experimental support, faced certain difficulties. The revolving electrons are constantly accelerated (centripetal) towards the nucleus. Such electrons, according to electromagnetic theory, would constantly radiate energy in the form of electromagnetic waves. Hence they would rapidly spiral in and fall into the nucleus, and the atom would collapse. (Calculations show that only about 10^{-10} sec would be required for a stable hydrogen atom to collapse). In practice, atom do not collapse.

Furthermore, in Rutherford's model the electrons can revolve in orbits of all possible radii and so they should emit continuous radiation of all frequencies. But the experimental fact is that atoms like hydrogen emit line spectra of radiation of only certain fixed frequencies.

From these difficulties we can only conclude that the classical laws of Physics which are true in macroscopic world do not hold in the microscopic world of the atom.

Contribution of Bohr

In 1913, Bohr gave a new and rather bold idea of atomic structure in order to explain the stability of the atom and the emission of sharp spectral lines. He proposed the following two modifications in Rutherford's model which are contrary to the classical theory :

1. The elections can revolve around the nucleus only in certain discrete orbits, and they do not radiate energy while in these permitted orbits. Hence the atom is stable. The permitted orbits are those in which the angular momentum *mvr** of the electron is an integral multiple of $\frac{h}{2\pi}$; *h* being the Planck's constant,

   i.e. $$\text{mvr} = \frac{nh}{2\pi}$$

   where $n = 1, 2, 3, \ldots\ldots$

2. The emission (or absorption) of radiation by the atom takes place when an electron jumps from one permitted orbit to another (Fig.). The radiation is emitted (or absorbed) as a single quantum (photon) whose energy hv is equal to the difference in energy ΔE of the electron in the two orbits involved. Thus

$$hv = \Delta E,$$

where v is the frequency of emitted (or absorbed) radiation. Hence the spectrum of the atom will have certain fixed frequencies.

The main objection against the Bohr's model was that its assumptions were entirely arbitrary, and could only be justified by the experimental results.

The Bohr's model as it is, was found lacking in explaining intensity of spectral lines, fine structure of spectral lines, spectra of complex atoms, etc.

Wave Mechanical Model: de Broglie gave the physical interpretation of Bohr's postulate that only certain orbits are permitted for the electron. He treated the electron (mass m, velocity v) as a wave of wavelength

$$\lambda = \frac{h}{mv},$$

which forms a stationary wave-pattern along the atomic orbits. He deduced that an electron can revolve around the nucleus indefinitely without radiating energy provided that its orbit contains an integral number of electron wavelengths (Fig.) *i.e.*

$$2\pi r = h\lambda.$$

But $$\lambda = \frac{h}{mv}$$

∴ $$2\pi r = \frac{h}{mv}$$

or $$mvr = n\frac{h}{2\pi},$$

which is same as the Bohr's postulate.

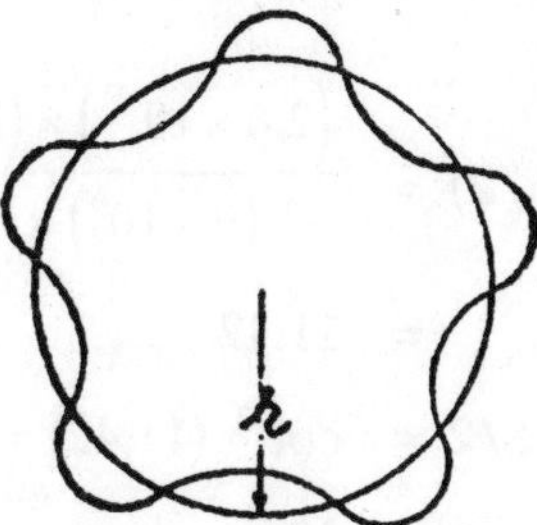

Later on, the application of quantum mechanics to the atom automatically yielded all the results of the Bohr theory and solved many problems which could not be handled by the Bohr theory. The fine structure of spectral lines was explained by considering the relativist variation of electron mass.

PROBLEMS

1. A 5-MeV α-particle approaches a gold (Z= 79) nucleus with an impact parameter of 2.6 × 10^{-15} meter. The angle through which it will be scattered is as follows (e = l.6 × 10^{-19} coulomb and 1 eV = l.6 × 10^{19} joule).

Solution : The angle of scattering ϕ of α-particle is related to the impact parameter p by

$$\cot(\phi/2) = \frac{2p}{b},$$

where $b = \frac{1}{4\pi\epsilon_0}\frac{2Ze^2}{E_k}$, E_k being the initial kinetic energy of the α-particle. Thus

$$\cot(\phi/2) = \frac{2p}{b} = \frac{(4\pi\epsilon_0)\,pE_k}{Ze^2}.$$

Here $p = 2.6 \times 10^{-13}$ meter, $Z = 79$; $e = 1.6 \times 10^{-19}$ coulomb and $E_k = 5$ MeV $= 5 \times 10^6$ eV $= 5 \times 10^6 \times (1.6 \times 10^{-19})$ joule.

Also $\frac{1}{4\pi\epsilon_0} = 9 \times 10^9$ nt-m^2/joul^2.

$$\therefore \qquad \cot(\phi/2) = \frac{\left(2.6 \times 10^{-13}\right) \times \left(5 \times 10^6 \times 1.6 \times 10^{-19}\right)}{\left(9 \times 10^9\right) \times 79 \times \left(1.6 \times 10^{-19}\right)^2}$$

$$= 11.42$$

$$\text{or} \qquad \phi/2 = \cot^{-1}(11.42) = 5^\circ$$

$$\therefore \qquad \phi = 10^\circ$$

2. *A gold foil of density 19.3 gm/cm^3 and atomic weight 197 gm/gm-atom has a thickness of 2.0×10^{-4} cm. It is used to scatter α-particles of kinetic energy 80 MeV. The fraction of the α-particles is scattered at angles greater than 90° is given below For gold Z = 79. (1 MeV=l.6 × l0^{13} J.) Avogadro's number is 6.02 × l0^{23} atoms/gm-atom.*

Solution : Let us first calculate the impact parameter p for scattering at 90°. We have

$$\cot\frac{\phi}{2} = \frac{2p}{b},$$

where $b = \frac{1}{4\pi\epsilon_0}\frac{2Ze^2}{E_k}$. From this,

$$p = \frac{1}{4\pi\epsilon_0}\frac{Ze^2}{E_k}\cot\frac{\phi}{2}$$

$$= (9 \times 10^9\ N\text{-}m^2/C^2)\,\frac{79 \times \left(1.6 \times 10^{-19}\ C\right)^2}{8.0 \times 1.6 \times 10^{-13}\ J}\cot 45^\circ$$

$$= 1.4 \times 10^{-14}\ m.$$

The cross-section is

$$\pi p^2/ = 3.14 \times (1.4 \times 10^{-14})^2$$

$$= 6.2 \times 10^{-28}\ \text{m}^2/\text{atom}.$$

The number of atoms per unit volume of the foil

$$n = \frac{\text{atoms}}{\text{volume}} = \frac{\text{mass/volume}}{\text{mass/atoms}}$$

$$= \frac{\text{density}}{\text{atomic wt/ Avogadro's number}}$$

$$= \frac{19.3\ \text{mg}/\text{cm}^3}{(197\ \text{gm}/\text{gm-atom})/(6.02\times 10^{23}\ \text{atom}/\text{gm-atom})}$$

$$= 5.9 \times 10^{22}\ \text{atoms/cm}^3$$

$$= 5.9 \times 10^{28}\ \text{atoms/m}^3.$$

Now, the fraction of the incident particles scattered through an angle greater than ϕ is given by

$$f_{>\phi} = \pi p^2\, nt,$$

where p is impact parameter for scattering angle ϕ and t is thickness of the foil. Thus

$$f_{>90^\circ} = (6.2\times 10^{-28}\ \text{m}^2/\text{atom})\ (5.9 \times 10^{28}\ \text{atoms/m}^3)\ (2.0 \times 10^{-6}\ m)$$

$$= 7.3 \times 10^{-5}.$$

3. A beam of α-particles of energy 5.5 MeV is incident normally on a gold (Z=79) foil at a rate of 10^4 particles/sec. An α-particle counter with an aperture 4 cm^2 is placed at a distance 0.25 in from the foil. The counts at an angle of 60° is calculated, if the thickness of the foil is 10^{-5} cm. The atomic weight (mass of 1 gm-atom) of gold is 197 and its density is 193 gm/cm^3. The Avogadro's number is 6.023 × 10^{23}.

Solution : The number of α-particles entering the counter per unit area at an angle of ϕ is given by

$$N(\phi) = \frac{N_0 nt\, Z^2 e^4}{4r^2 (4\pi\epsilon_0)^2\, E_k^2}\ \frac{1}{\sin^4(\phi/2)}$$

Here, N_0 (particles striking the foil) = 10^4/sec,

n (number of atoms in the foil per unit volume)

$$= \frac{\text{mass / volume}}{\text{mass / no. of atoms}}$$

$$= \frac{\text{density}}{\text{atomic wt / Avogadro's number}}$$

$$= \frac{19.3 \text{ gm / cm}^3}{197 \text{ gm / } 6.023 \ 10^{23}}$$

$$= 5.9\ 5\ 10^{22}/\text{cm}^3 = 5.9 \times 10^{28}/\text{m}^3,$$

t (thickness of foil) = 10^{-5} cm = 10^{-7} m,

r (distance of counter from the foil) = 0.25 m,

Z = 79, $e = 1.6 \times 10^{-19}$ coul,

E_k = 5.5 MeV=5.5 × 1.6 × 10^{-13}

joule and $\phi = 60°$ so that $\sin \phi/2 = \sin 30° = \frac{1}{2}$. Thus

$$N\ \ (\phi) = \frac{10^4 \times (5.9 \times 10^{28}) \times 10^{-7} \times (79)^2 \times (1.6 \times 10^{-19})^4}{4 \times (0.25)^2 \times \left(\frac{1}{9 \times 10^9}\right)^2 \times (5.5 \times 1.6 \times 10^{-13})^2} \times \frac{1}{\left(\frac{1}{2}\right)^2}$$

$$= \frac{5.9 \times (79)^2 \times (1.6)^2 \times (9)^2 \times (2)^4}{4 \times (0.25)^2 \times (5.5)^2} \times 10^{-7} = 1.6.$$

The area of the aperture of the counter is 4 cm^2 = 4 ×10^{-4} m^2. Therefore the number of particles entering the counter

$$= 1.6 \times (4 \times 10^{-4}) = 6.4 \times 10^{-4}/\text{sec.}$$

4. In Rutherford scattering experiment the number of particles observed at an angle of 10° is one million/rain, the number of particles per minute will be observed at 90° and at 180° is explained below.

Solution : By Rutherford formula, the number of particles of a given energy scattered at angle ϕ from a given foil is

$$N(\phi) \propto \frac{1}{\sin^4(\phi/2)} = K\frac{1}{\sin^4(\phi/2)}$$

where K is constant. Now $N(10°) = 10^6$/min.

Thus $$10^5 = K\frac{1}{\sin^4\left(\frac{10}{2}\right)^\circ} = \frac{K}{(0.0872)^4}$$

so that $$K = (0.0872)^4 \times 10^6.$$

Hence the particles observed at 90° are

$$N(90°) = K\frac{1}{\sin^4\left(\frac{90}{2}\right)^\circ}$$

$$= (0.0872)^4 \times 10^6 \times \frac{1}{\left[\left(1/\sqrt{2}\right)\right]^4}$$

$$= 232.$$

Similarly, $N(180°) = 58.$

5. *The distance of closest approach of α-particles to the copper nucleus, when α-particles of 5 MeV energy are scattered back by a thin sheet of copper (Z = 29) is calculated as follows.*

Solution : An α-particle will have its closest approach when its impact parameter is zero, *i.e.*, when it moves directly (head-on) toward the nucleus and is *reflected* back (180° scattering).

Let r_0 be the distance of the α-particle from the nucleus at the closest approach. The (positive) charge on the nucleus is Ze, and that on the α-particle is $2e$, where e is the electronic charge. Therefore, the electrostatic potential energy of the particle at the instant of closest approach is $\frac{1}{4\pi\epsilon_0}\frac{(Ze)(2e)}{r_0}$

At this instant the α-particle is momentarily at rest, and its initial kinetic energy E_k is entirely converted into electrostatic potential energy. Hence at this instant ; we have

$$E_k = \frac{1}{4\pi\epsilon_0}\frac{2Ze^2}{r_0}.$$

$$\therefore \quad r_0 = \frac{1}{4\pi\epsilon_0}\frac{2Ze^2}{E_k}$$

Here $E_k = 5$ MeV $= 5 \times 10^6$ eV $= (5 \times 10^6) \times (1.6 \times 10^{-19}$ joule);

$Z = 29$ and $e = 1.6 \times 10^{-19}$ coul. Also $\frac{1}{4\pi\epsilon_0} = 9 \times 10^9$ nt-m^2/coul^2.

$$\therefore \quad r_0 = 9 \times 10 \times \frac{2 \times 29 \times (1.6 \times 10^{-19})^2}{5 \times 10^6 \times 1.6 \times 10^{-19}}$$

$$= 1.67 \times 10^{-14} \text{ meter.}$$

(*b*) The (positive) charge on the proton is *e* (*i.e.* half of that on α-particle). Therefore the electrostatic potential energy at the instant of closest approach is $\frac{1}{4\pi\epsilon_0}\frac{(Ze)(2e)}{r_0}$. Thus, as above

$$E_k = \frac{1}{4\pi\epsilon_0}\frac{Ze^2}{r_0}$$

$$\therefore \quad r_0 = \frac{1}{4\pi\epsilon_0}\frac{Ze^2}{E_k}$$

which is half of that of an α-particle of the same energy. Hence for the proton, we have

$$r_0 = \frac{1.67 \times 10^{-14}}{2} = 0.835 \times 10^{-14} \text{ meter.}$$

When an energetic α-particle moves directly (head-on) towards a nucleus, it experiences an increasing electrostatic force of repulsion in accordance to the inverse-square law. Hence its velocity goes on decreasing and on reaching a certain minimum distance from the nucleus the particle comes momentarily to rest. It is then accelerated backwards, reverses its motion and is said to be scattered through 180°. The minimum distance r_0 to which the incident α-particle approaches the nucleus is called the 'distance of closest approach, and is related to the initial kinetic energy E_k of the α-particle by

$$r_0 = \frac{1}{4\pi\epsilon_0}\frac{2Ze^2}{E_k}.$$

The greater the initial kinetic energy E_k of the particle, the smaller the distance of closest approach r_0 . If, however, the energy E_k exceeds a certain maximum value E, the α-particle approaches the nucleus so much closely that the nuclear size becomes comparable with the distance of the particle from it and the inverse-square law breaks down. Besides this, some short-range *attractive* nuclear forces also begin to operate. Hence under this condition the α-particle is not scattered back. The distance of closest approach of a particle of maximum kinetic energy E gives the upper limit of the radius of the nucleus. Thus if the maximum nuclear radius be R, we have

$$R = \frac{1}{4\pi\epsilon_0}\frac{2Ze^2}{E}$$

or $$E = \frac{1}{4\pi\epsilon_0}\frac{2Ze^2}{R}.$$

Here $Z = 79$; $e = 1.6 \times 10^{-19}$ coul

and $R = 0.5$ Å $= 0.5 \times 10^{-10}$ meter.

$$\therefore \quad E = (9 \times 10^{-9}) \times \frac{2 \times 79 \times \left(1.6 \times 10^{-19}\right)^2}{0.5 \times 10^{-10}}$$

$$= 7.3 \times 10^{-16} \text{ joule.}$$

Speed of Fastest aα-particle suffering 180°-Scattering: Let E_k be the kinetic energy of the fastest α-particles. The distance of the closest approach of the most energetic (fastest) a-particles suffering 180°-scattering from a nucleus is the maximum possible radius of that nucleus. If the maximum radius of the nucleus (charge + Ze) is R, then the electrostatic potential energy of the α-particle (charge $+2e$) at the instant of closest approach will be $\frac{1}{4\pi\epsilon_0}\frac{2Ze^2}{R}$. Since at this instant the entire kinetic energy E_k has been converted into the potential energy; we have

$$E_k = \frac{1}{4\pi_0},\ \frac{2Ze^2}{R}$$

But $E_k = \frac{1}{2} m_\alpha v^2$, where m_α is the mass and v the speed of the fastest particle. Also, given that $m_\alpha = 4m_y$, where m_p is the mass of a proton. Thus

$$E_k = \frac{1}{2}(4m_p)\,v^2 = \frac{1}{4\pi\epsilon_0}\frac{2Ze^2}{R}$$

or

$$v^2 = \frac{1}{4\pi\epsilon_0}\frac{Ze^2}{m_p R}$$

∴

$$v = \sqrt{\left(\frac{Ze^2}{(4\pi\epsilon_0)\,m_p R}\right)}$$

An α-particle is scattered back (180°) at its closest approach to the nucleus. The distance of closest approach is the upper limit of the nuclear radius. If it is R, then we have

$$R = \frac{1}{4\pi\epsilon_0}\frac{2Ze^2}{E_k}$$

$$= \left(9\times10^9\ \frac{N-m^2}{C^2}\right)\frac{2\,(79)\left(1.6\times10^{-19}\ C\right)^2}{4\times1.6\times10^{-13}\ J}$$

$$= 5.7\times10^{-14}\ \text{m}.$$

∴ maximum volume in which the positive charge may be concentrated

$$= \frac{4}{3}\pi R^3$$

$$= \frac{4}{3}(3.14)\,(5.7\times10^{-14}\ \text{m})^3 = 7.75\times10^{-40}\ \text{m}^3.$$

6. A beam of protons and α-particles, each of energy 15 MeV is incident on a thin foil of tin (Z = 50). It is observed that the α-particles are scattered backward but not the protons. This observation explained and the maximum possible value of the nuclear radius is found. As to why electrons are not effective in scattering α-particles and protons are also explained.

Solution : The distance of closest approach of α-particle (charge + 2*e)* to a nucleus (charge +Ze) is given by

$$r_0 = \frac{1}{4\pi\epsilon_0}\frac{2Ze^2}{E_k},$$

where E_k is the initial kinetic energy of the incident α-particle. The particle on reaching at this distance from the nucleus is reflected back (scattered 180°).

Here $E_k = 15\text{ MeV} = 15\times 10^6\text{ eV} = 15\times 10^6 \times 1.6 \times 10^{-19}$ joule.

$$\therefore \quad r_0 = (9 \times 10^9) \times \frac{2\times 50\times\left(1.6\times 10^{19}\right)^2}{15\times 10^6\times 1.6\times 10^{-19}}$$

$$= 9.6 \times 10^{-15} \text{ meter.}$$

If 15 MeV is the maximum energy of the α-particles for which they are scattered backwards, then 9.6×10^{-15} meter is also the maximum possible value of the radius of the tin nucleus.

The charge on a proton is + *e, i.e.* half of that on an α-particle. Hence a proton of 15 MeV energy will have its distance of closest approach half of that of α-particle *i.e.* 4.8×10^{-15} meter. This means that the proton of this energy will penetrate the nucleus, and not be scattered backward.

The electrons, because of their little mass, are unable to deviate the α-particles and protons appreciably.

7. With gold foil (Z=79), 180-scattering is observed with α-particles of velocity upto 16 × 10^7 m/sec. The upper limit for

radius of the nucleus is deduced. The mass of an α-particle is $4m_p$ where m_p is the mass of proton (1.67×10^{-27} kg).

Solution : The distance of closest approach of an α-particle suffering 180° scattering from a nucleus of atomic number Z; given by

$$r_0 = \frac{1}{4\pi\epsilon_0}\frac{2Ze^2}{E_k}$$

where E_k is the initial kinetic energy of the particle. If m_α be the mass and v the velocity of the particle, then $E_k = \frac{1}{2} m_\alpha v^2$, so that

$$r_0 = \frac{1}{4\pi\epsilon_0}\frac{4Ze^2}{m_\alpha v^2}.$$

If v be the velocity of the *fastest* particle that can suffer 180° scattering, then r_0 is also the upper limit for the radius of the nucleus.

Putting the given values; the upper limit of gold nucleus is

$$r_0 = \frac{\left(9 \times 10^9\right) \times 4 \times 79 \times \left(1.6 \times 10^{-19}\right)^2}{\left(4 \times 1.67 \times 10^{-27}\right) \times \left(1.6 \times 10^7\right)^2}$$

$$\approx 4.26 \times 10^{-14} \text{ meter.}$$

2

Nuclear Reactions

Damages of Nuclear Reactions

A nuclear reaction is a strong interaction of an atomic *nucleus* with an elementary particle, which results in a new nucleus and one or more new particles. The reaction occurs when the interacting nucleus and the particle approach each other up to distances of the order of 10^{-15} meter.

An equation representing a nuclear reaction can be written as

$$X + a \rightarrow Y + b,$$

where X is the initial nucleus, a the initial particle, Y the final nucleus and b the final particle. It can be written in the following abbreviated form :

$$X\ (a, b)\ Y$$

The particles participating in the reaction are indicated in parentheses, first the initial particle, then the final one.

The particles a and b may be alpha particle ($_2He^4$), proton ($_1H^1$ or p), deuteron ($_1H^2$ or d), neutron (n), gamma-ray photon (γ), etc.

Nuclear reactions are usually attended by the liberation or absorption of energy, which is called the 'reaction energy' or 'disintegration energy'.

The first nuclear reaction discovered by Rutherford is

$$_7N^{14} + {}_2He^4 \rightarrow {}_8O^{17} + {}_1H^1$$

(α-particle) (proton)

or $_7N^{14}\ (\alpha,\ p)\ {}_8O^{17}$.

In this reaction a nitrogen nucleus hit by an α-particle is converted into an oxygen nucleus and emits a proton. The reaction is called an alpha-proton (α, p) reaction.

Like a chemical reaction, a nuclear reaction must be balanced; the number of protons and neutrons must be the same on both sides of the equation.

Theory of Compound Nucleus

Bohr, in 1936, proposed the compound nucleus hypothesis for nuclear reactions. According to it, many nuclear reactions proceed in two separate stages :

(i) The incident particle strikes the initial (or target) nucleus and the two first combine to form a new nucleus, called a *compound nucleus.* The atomic and mass numbers of the compound nucleus are respectively the sum of the atomic numbers and the sum of the mass numbers of the incident particle and the target nucleus. The energy brought by the incident particle is quickly shared by the nucleons of the compound nucleus which is therefore invariably in an excited state.

(ii) The (excited) compound nucleus immediately breaks up by ejecting a particle or a γ–photon, leaving the final (or product) nucleus.

Thus the Rutherford's reaction can be written to show the intermediate stage of the compound nucleus formation as

$$_7N^{14} + {}_2He^4 \rightarrow ({}_9F^{18}) \rightarrow {}_8O^{17} + {}_1H^1.$$

The compound nucleus has an average life-time of the order of 10^{-14} sec which is very long compared with the time ($\sim 10^{-21}$ sec) required for a particle with an energy of several MeV to pass through a nucleus. The compound nucleus therefore "forgets" how it was formed and its breaking is independent of the mode of its formation. In fact, the same compound nucleus can be formed by *different* pairs of incident particle and target nucleus, and can break in different ways, giving different products. The idea of compound nucleus has been very fruitful in correlating many nuclear phenomena.

Stripping Reaction and Pick-up Reaction : A different type of nuclear reaction is sometimes observed when a deuteron strikes a nucleus. Because of the finite distance between the proton and the neutron in the deuteron, one of these particles reaches the nuclear surface earlier and is captured by the nucleus, whereas the other particle flies past the nucleus. This is called 'stripping reaction' because a nucleon is stripped out of the deuteron by the target nucleus.

The reverse of a stripping reaction is a 'pick-up reaction'. A nucleon (*n or p*) striking a nucleus picks-up from it a nucleon (*p* or *n*) and forms a deuteron.

Conservation Laws in Nuclear Reactions : The nuclear reactions are governed by certain conservation laws which limit the possible outcomes of the reaction. Some of these laws are as follows :

1. *Conservation of Total Relativistic Energy* **:** Mass and kinetic energy are *not* separately conserved in nuclear reactions. Instead, the total relativistic energy (kinetic energy + rest mass energy) is conserved.

Let us consider the general case in which a bombarding particle a interacts with a target nucleus X to give a product nucleus Y and a product particle b ; that is

$$X + a \rightarrow Y + b.$$

According to the conservation law, the kinetic energy plus the rest energy on one side of the reaction equation must equal the corresponding sum on the other side ; that is

$$m_x c^2 + (K_a + m_a c^2) = (K_y + m_y c^2) + (K_b + m_b . c^2).$$

Here

$K_x = 0$ since X is stationary in the laboratory frame.

2. *Conservation of Linear Momentum* : The resultant linear momentum of the system $(X + a)$ must be equal in magnitude and direction to that of the system $(Y + b)$. It helps in determining the energy balance of the nuclear reaction.

3. *Conservation of Angular Momentum:* The sum of the spin angular momentum and the motional angular momentum of the system $(X+a)$ should be equal to the corresponding sum of the system $(Y + b)$. The angular momentum affects the rate, or cross-section, of the nuclear reaction.

4. *Conservation of Charge :* The total net electric charge before and after the nuclear reaction should be the same.

5. *Conservation of the number of Nucleons (Mass Number):* In all nuclear reactions the number of nucleons (or the mass number) remains conserved, *i.e.*, the total number of nucleons present before the reaction equals the total number present after.

6. *Conservation of Parity :* Parity is also conserved in a nuclear reaction.

Nuclear Reactions induced by α-particles : The (positively-charged) α-particles can readily penetrate light (low Z) nuclei which offer less coulombic repulsion. Besides the Rutherford's first reaction, some other reactions induced in *light* nuclei by α-particles with energies upto about 7-8 MeV emitted from natural substances are as follows :

$$\left.\begin{array}{l} {}_5B^{10} + {}_2He^4 \rightarrow ({}_7N^{14}) \rightarrow {}_6C^{13} + {}_1H^1 \\ {}_{12}Al^{27} + {}_2He^4 \rightarrow ({}_{15}P^{31}) \rightarrow {}_{14}Si^{30} + {}_1H^1 \end{array}\right\} \quad (\alpha,\ p) \text{ reactions}$$

$$\left.\begin{array}{l} {}_4Be^9 + {}_2He^4 \rightarrow ({}_6C^{13}) \rightarrow {}_6C^{12} + {}_0n^1 \\ {}_5B^{11} + {}_2He^4 \rightarrow ({}_7N^{15}) \rightarrow {}_7N^{14} + {}_0n^1 \\ {}_7N^{14} + {}_2He^4 \rightarrow +({}_9F^{18}) \rightarrow {}_9F^{17} + {}_0n^1 \\ {}_{13}Al^{27} + {}_2He^4 \rightarrow ({}_{15}P^{31}) \rightarrow {}_{15}P^{30} + {}_0n^1 . \end{array}\right\} \quad (\alpha,\ n) \text{ reactions}$$

As we are seeing, a given incident particle and a given target nucleus can lead to different reaction products depending on the energy of the incident particle.

α-particles directed towards *heavier* nuclei experience strong coulombic repulsion and are scattered rather than reacting with the nuclei.

Reactions induced by Protons : Protons (and deuterons) with a single positive charge would experience smaller coulombic repulsion than α-particles in the field of the nucleus, Hence these particles, when accelerated to high energies by means of particle-accelerators can induce reactions in many nuclei. Cockcroft and Walton were the first to produce such an artificially induced reaction in 1932. They disintegrated lithium-7 by protons accelerated, to about 0.3 MeV which resulted in α-particle production :

$${}_3Li^7 + {}_1H^1 \rightarrow ({}_4Be^8) \rightarrow {}_2He^4 + {}_2He^4 \qquad (p,\ 2\alpha)$$

This reaction had provided one of the earliest verifications of the Einstein's mass-energy formula. The other proton-induced reactions which give rise to α-particles, neutrons, and γ–rays are as follows:

$$\left.\begin{array}{l} {}_4Be^9 + {}_1He^1 \rightarrow ({}_5B^{10}) \rightarrow {}_3Li^6 + {}_2He^4 \\ {}_9F^{19} + {}_1He^1 \rightarrow ({}_{10}Ne^{20}) \rightarrow {}_8O^{16} + {}_2H^4 \\ {}_{13}Al^{27} + {}_1H^1 \rightarrow ({}_{14}Si^{28}) \rightarrow {}_{12}Mg^{24} + {}_2He^4 \end{array}\right\} \quad (p,\ \alpha)$$

$${}_5B^{11} + {}_1H^1 \rightarrow ({}_6C^{12}) \rightarrow {}_2He^4 + {}_2He^4 + {}_2He^4 \qquad (p,\ 3\alpha)$$

$$_8O^{18} + _1H^1 \rightarrow (_9F^{19}) \rightarrow {}_9F^{18} + {}_0n^1 \qquad (p, n)$$

$$_3Li^7 + _1H^1 \rightarrow (_4Be^8) \rightarrow {}_4Be^8 + \gamma \qquad (p, \gamma)$$

In the last reaction the bombarding proton is captured by the target nucleus forming an excited compound nucleus which is simply de-excited by emitting γ–ray photon. Such a reaction is called 'radiative capture'.

Reactions induced by Denterons : Many nuclear reactions have been observed with deuterons accelerated to energies over a wide range upto several hundred MeV, producing α-particles, protons and neutrons.

$$_2Li^6 + _1H^2 \rightarrow (_4Be^8) \rightarrow {}_2He^4 + {}_2He^4 \qquad (d, 2\alpha)$$

$$_8O^{16} + _1H^2 \rightarrow (_9F^{18}) \rightarrow {}_7N^{14} + {}_2He^4 \qquad (d, \alpha)$$

$$_6C^{12} + _1H^2 \rightarrow (_7N^{14}) \rightarrow {}_6C^{13} + {}_1H^1 \qquad (d, p)$$

$$_6C^{12} + _1H^2 \rightarrow (_7N^{14}) \rightarrow {}_7n^{13} + {}_0n^1 \qquad (d, n)$$

Reactions induced by Neutrons : Neutrons, being uncharged, are much more likely to penetrate nuclei than are protons, deuterons or α-particles. Therefore fast as well as slow neutrons have been found extremely effective in producing nuclear disintegrations.

The reaction between a neutron and a nucleus gives rise in most cases to an α-particle, a proton, two neutrons, or a γ–ray photon (radiative capture). The actual type of reaction depends upon the energy of the bombarding neutron and the nature of the target nucleus. Among the most interesting neutron-induced reactions are the following:

$$\left.\begin{array}{l} _3Li^6 + {}_0n^1 \rightarrow (_3Li^7) \rightarrow {}_1H^3 + {}_2He^4 \\ _5B^{10} + {}_0n^1 \rightarrow (_5B^{11}) \rightarrow {}_3Li^7 + {}_2He^4 \\ _{13}Al^{27} + {}_0n^1 \rightarrow (_{13}Al^{28}) \rightarrow {}_{11}Na^{24} + {}_2He^4 \end{array}\right\} \qquad (n, \alpha)$$

$$\left.\begin{array}{l} _7N^{14} + {}_0n^1 \rightarrow (_7N^{15}) \rightarrow {}_6C^{14} + {}_1He^1 \\ _{13}Al^{27} + {}_0n^1 \rightarrow (_{13}Al^{28}) \rightarrow {}_{12}Mg^{27} + {}_1H^1 \end{array}\right\} \qquad (n, p)$$

$$_{13}Al^{27} + _0n^1 \rightarrow (_{13}Al^{28}) \rightarrow _{13}Al^{26} + _0n^1 + _0n^1 \qquad (n, p)$$

$$\left.\begin{array}{l} _1H^1 + _0n^1 \rightarrow (_1H^2) \rightarrow _1H^2 + \gamma \\ _{13}Al^{27} + _0n^1 \rightarrow (_{13}Al^{28}) \rightarrow _{13}Al^{28} + \gamma \\ _{92}U^{238} + _0n^1 \rightarrow (_{92}U^{239}) \rightarrow _{92}U^{239} + \gamma \end{array}\right\} \text{radiative } (n, \gamma) \text{ capture}$$

The *slow* neutrons, almost in all cases, induce (n, γ) reactions (radiative capture) and the product nuclei are often radioactive. These reactions are important source of artificial radio-isotopes.

Reactions induced by Photons (Photo disintegration) : Nuclei can also be disintegrated by bombardment with γ–ray photons, provided that the energy of the photons exceeds a threshold value depending on the target nucleus. For example, deuteron can be disintegrated by bombarding with 2.62-MeV γ–rays emitted from ThC. The reaction is

$$_1H^2 + \gamma \rightarrow (_1H^2) \rightarrow _1H^1 + _0n^1$$

for which the threshold energy is 2.2 MeV. This is the binding energy of the proton and the neutron in the deuteron nucleus. Reactions like (γ, 2n), (γ, αn), (γ, p) have been observed by using high-energy γ–ray photons derived from a betatron.

Q-value of Nuclear Reaction : In a nuclear reaction not only the charge numbers and mass numbers balance, but also the total relativistic energy, (= kinetic energy + rest mass energy) remains conserved. Let us consider a general nuclear reaction

$$X + a = Y + b, \qquad \text{... } (i)$$

where X is the target nucleus, a is the bombarding particle, Y is the product nucleus, and b is the product particle.

Let m_x , m_a , m_y and m_b be the *rest* masses of X, a, Y and b respectively; and K_a , K_y and K_b the kinetic energies of a, Y and b respectively. The target nucleus X is initially at rest so that it has no kinetic energy ($K_x = 0$). By the conservation of *total* relativistic energy in reaction (*i*), we have

$$m_x c^2 + (K_{a'} + m_a c^2) = (K_y + my\ c^2) + (K_b + M_b\ c^2). \quad ...\ (ii)$$

Because there can be an exchange of energy between kinetic energy and rest mass energy, the final kinetic energy $K_y + K_b$ may be greater, or less, than the initial kinetic energy K_a. The difference is called the '*Q*-value' or 'energy balance' of the reaction, That

$$Q = K_y + K_b - K_a. \quad ...\ (iii)$$

Substituting the value of K_y+K_b—K_a from eq. (*ii*), we obtain

$$Q = (m_X + m_a - m_y - m_b)\ c^2 \quad ...\ (iv)$$

Thus the *Q*-value may be determined by measuring the energies or the masses.

The m's in eq. (iv) refer to the rest masses of the *nuclei* X *and* Y and to the rest masses of the completely ionised *nuclear* particles *a* and *b*. However, in view of the conservation of charge in nuclear reaction, we can incorporate electron masses in m's (ignoring electron binding energy) and hence can treat them as *atomic* masses.

Reactions for which *Q* is positive, the rest mass energy of product particles *Y* and *b* is less than that of the initial particles *X* and *a*. The rest energy surplus of X and *a* is converted into the kinetic energy of the product particles which exceeds the kinetic energy of the initial particles. Such reactions are called 'exothermic' or 'exoergic'.

If *Q* is negative, that is, the rest energy of the product particles exceeds that of the initial particles, the reaction is 'endothermic' or 'endoergic'. In this case the surplus rest energy of the product particles results in a loss of kinetic energy.

Meaning of Atomic Mass Unit

Atomic mass unit (abbreviated 'u') is the unit in which atomic masses are expressed. "l u is defined as 1/12 the mass of an atom of $_6C^{12}$ which is the most abundant isotope of carbon". Thus, by definition, the mass of $_6C^{12}$ *atom* is exactly

12.00000......*u*. The masses of other atoms are determined by measuring the position of the line they produce in a mass spectrograph relative to that of $_6C^{12}$.

Atomic masses refer to the masses of neutral atoms, not of stripped nuclei. Thus the masses of the orbital electrons and the mass equivalent of their binding energies. (which are very small) are included in the figure of atomic masses.

1 gm-atom of $_6C^{12}$ has *N* (Avogadro's number) atoms and a mass of 12 gm. Therefore, the mass of 1 atom of $_6C^{12}$ is $\frac{12}{N}$ gm.

Thus, by definition.

$$1 \text{ u} = \frac{1}{12} \text{ mass of 1 atom of } {}_6C^{12}$$

$$= \frac{1}{12}\left(\frac{12}{N} \text{gm}\right) = \frac{1}{N} \text{ gm.}$$

But $$N = 6.023 \times 10^{23}.$$

$\therefore$ $$1 \text{ u} = \frac{1}{6.023 \times 10^{23}} = 1.66 \times 10^{-24} \text{ gm}$$

or $$\boxed{1\, u = 1.66 \times 10^{-27} \text{ kg.}}$$

Let us now calculate the energy equivalence of 1 u. We know that 1 u = 1.66×10^{-27} kg and $E = mc^2$. Thus, for 1 u, we have

$$E = 1.66 \times 10^{-27} \text{ kg} \times (3.0 \times 10^8 \text{ m/s})^2$$

$$= 1.49 \times 10^{-10} \text{ joule.}$$

But 1 eV = 1.6×10^{-19} joule.

$\therefore$ $$E = \frac{1.49 \times 10^{-10} \text{ joule}}{1.6 \times 10^{-19} \text{ joule / eV}}$$

$$= 0.931 \times 10^9 \text{ eV} = 931 \text{ MeV.}$$

Using more accurate values of the constants involved yields 931.5 MeV. Thus

$$1\,u \equiv 931.5\text{ MeV}$$
$$1\,u \times c^2 = 931.5\text{ MeV}$$

PROBLEMS

1. ***Complete the following nuclear reactions :***

(i) $_{17}Cl^{35} + ? \rightarrow {}_{16}S^{32} + {}_2He^4$

(ii) $_5B^{10} + ? \rightarrow {}_3Li^7 + {}_2He^4$

(iii) $_3Li^6 + ? \rightarrow {}_4Be^7 + {}_0n^1$

(iv) $_{13}Al^{27} + {}_0n^1 \rightarrow {}_{12}Mg^{27} + ?$

(v) $_4Be^9 + {}_2He^4 \rightarrow ? + {}_0n^1$

(vi) $_3Li^7$ + proton → α-particle + ?

(vii) deuteron+deuteron → proton + ?

[**Ans.** (*i*), $_1H^1$, (*ii*) $_0n^1$, (*iii*) $_1H^2$, (*iv*) $_1H^1$, (*v*) $_6C^{12}$, (*vi*) α-particle $_2He^4$, (*vii*) tritron $_1H^3$]

2. ***Complete the following disintegration reactions by substituting the proper nuclide or particle for the question mark in each case :***

(i) $_{14}Si^{28}$ (n,p) ?

(ii) $_3Li^7$ (p, γ) ?

(iii) $_{13}Al^{27}$ (p, γ) ?

(iv) $_{11}Na^{23}$ (p ; ?) $_{10}Ne^{20}$

(v) $_{13}A1^{27}$ (?, α) $_{11}Na^{24}$

(vi) $_5B^{10}$ (? , a) $_3Li^7$

(vii) $_{11}Na^{23}$ (d , ?) $_{12}Mg^{24}$

[**Ans.** (*i*) $_{13}Al^{28}$, (*ii*) $_4Be^8$, (*iii*) $_{14}Si^{30}$, (*iv*) $_2He^4$, (*v*) $_0n^1$ (*vi*) $_0n^1$, (*vii*) $_0n^1$.]

3. ***Complete the following reactions :***

(i) H^1 ***(n , γ) ?, (ii)*** Li^7 ***(p , ?)*** Be^7***, (iii)*** N^{14} ***(? , p)*** O^{17}***,***

(iv) ? (n , p) Na^{24}***.***

[Ans. ${}_1H^2$, ${}_0n^1$, ${}_2He^4$, ${}_{12}Mg^{24}$**].**

4. ***Calculate the energy released (Q-value) in the following reaction :***

$$ {}_3Li^6 + {}_1H^2 \rightarrow {}_2He^4 + {}_2He^4 + Q. $$

The atomic masses are : ${}_1H^2 = 2.014102\ u$, ${}_3Li^6 = 6.015123\ u$, ${}_2He^4 = 4002603\ u$***. Given*** $lu \times c^2 = 931.5$ ***MeV.***

Solution: The Q-value is initial masses minus final masses, multiplied by c^2, that is,

$$ Q = [m({}_3Li^6) + m\ ({}_1H^2) - m({}_2He^4) - m({}_2H^4)]c^2. $$

Putting the given values in atomic mass unit (u), we have

$$ Q = (6.015123\ u + 2.014102\ u - 4002603\ u - 4.002603\ u)c^2 $$

$$ = (0.024019\ u)c^2. $$

$$ \text{But } 1\ u \times c^2 = 931.5\ \text{MeV}. $$

$$ \therefore \quad Q = 0.024019 \times 931.5 $$

$$ = 22.37\ \text{MeV}. $$

5. ***Compute the Q-value of the reaction*** ${}_1H^3$ ***(d, n)*** ${}_2He^4$***. Given : mass of neutron = 1.008665 u, mass of deuterium (atom) =2.014102 u, mass of tritium (atom)=3.016049 a, mass of helium atom = 4.002603 a and 1 amu ≡ 931.5 MeV.***

Solution: The reaction is

$$ {}_1H^3 + {}_1H^2 \rightarrow {}_2He^4 + {}_0n^1. $$

Its Q value is given by

$$ Q = [m({}_1H^3) + m({}_1H^2) - m({}_2He^4) - m({}_0n^1)]c^2 $$

$$= (3.016049 \text{ u} + 2.014102\, u - 4.002603 \text{ u} - 1.008665 \text{ u})c^s$$

$$= (0.018883 \text{ u})\, c^2$$

$$= 0.018883 \times 931.5 = 17.6 \text{ MeV}.$$

6. *Compute the Q-value of the Rutherford's reaction :*

$${}_7N^{14} + {}_2He^4 \rightarrow {}_8O^{17} + {}_1H^1.$$

The atomic masses are : $_7N^{14}$ ***= 14.003074 u,*** $_2He^4$ ***= 4.002603 u,*** $_8O^{17}$ ***= 16.99913l u and*** $_1H^1$ ***= 1.007825 u.***

Solution: The Q – value of the reaction is given by

$$Q = [\text{m}({}_7\text{N}^{14}) + m\,({}_2\text{He}^4) - m({}_8\text{O}^{17}) - m({}_1\text{H}^1)]c^2$$

$$= (14.003074 \text{ u} + 4.002603\, u - 16.999131 \text{ u} - 1.007825 \text{ u})\, c^2$$

$$= 18.005677\, u - 18.006956\, u)\, c^2$$

$$= (-0.001279\, u)\, c^2.$$

The energy equivalent of mass 1 u is 931.5 MeV, that is, lu × c^2 = 931.5 MeV.

$$\therefore \qquad Q = -\,0.001279 \times 931.5 = -\,1.19 \text{ MeV}.$$

Because the Q-value is negative, the reaction is "endothermic"

7. *Compute the Q-value of the Chadwick's reaction :*

$${}_4Be^9 + {}_2He^4 \rightarrow {}_6C^{12} + {}_0n^1.$$

The atomic masses are : $_4Be^9$ ***= 9.012182 u,*** $_2He^4$ ***= 4.002603 u,*** $_6C^{12}$ ***= 12.000000 u, neutron mass = 1. 008665 u and 1 amu is equivalent to 931.5 MeV of energy.***

[**Ans.** 5.70 MeV.]

8. *Find the mass of* $_6C^{14}$ ***(in amu) from the following nuclear reaction :***

$$_7N^{14} + {}_0n^1 = {}_6C^{14} + {}_1H^1 + 0.55 \text{ MeV}.$$

Given : mass of proton = 1.00758 amu, mass of neutron = 1.00898 amu, mass of $_7N^{14}$ = 14.00752 amu and 1 amu = 931 MeV.

Solution: The *Q*-value of the reaction is 0.55 MeV. We know that

$$Q = [m({}_7N^{14}) + m({}_0n^1) - m({}_6C^{14}) - m({}_1H^1)]\,c^2.$$

Let the mass of $_6C^{14}$ be *mc* amu. Substituting the given values in the above equation, we have

$$(14.00752 \text{ amu} + 1.00898 \text{ amu} - mc - 1.00758 \text{ amu})\,c^2$$

$$= 0.55 \text{ MeV}$$

or$(14.00892 \text{ amu} - mc)c^2 = 0.55 \text{ MeV}.$

Now 1 amu × c^2 = 931 MeV. Therefore, we can write

$$(14.00892 \text{ amu-}mc)c^2 = \frac{1 \text{ amu} \times c^2}{931} \times 0.55$$

or$14.00892 \text{ amu} - mc = 0.00059 \text{ amu}$

or $mc = 14.00892 - 0.00059 = 14.00833$ amu.

9. ***Calculate the minimum energy of a photon that will break the nuclide $_3Li^7$ into an α-particle and a tritron. The atomic masses are : $_8Li^7$ = 7.016004 u, $_2He^4$ = 4.002603 u, $_1H^8$ = 3.016050 u and 1 u × c^2 = 931.5 MeV.***

Solution: The equation for the given photo-disintegration process is

$$_3Li^7 + \gamma = {}_2He^4 + {}_1H^3$$

(photon) (α-particle) (tritron)

The *minimum* energy of the photon must be equal to the *rest* energy of the product particles minus the (rest) energy of $_8Li^7$. Let us write the above equation in terms of energy :

$$\gamma = [m({}_2He^4) + m\,({}_1He^3) - m({}_3Li^7)]c^2$$
$$= (4.002603 \text{ u} + 3.016050 \text{ u} - 7.016004 \text{ u})\ c^2$$
$$= (0.002649 \text{ u})c^2$$
$$= 0.002649 \times 931.5 = 2.47 \text{ MeV}.$$

10. ***Complete the following reactions and calculate the Q-value in each case. ($_4Be^9$ = 9012182 u. $_4Be^7$ = 7.016930 u, $_3Li^7$ = 7.016004 u, $_3Li^6$ = 6.015123 u, $_1H^1$ 1.007825 u, $_1He^4$ = 4.002603 u, $_0n^1$=1-0.08665 u, 1 u × c^2 = 931.5 MeV)***

(i) $_3Li^7$ ***(p, n) ?***

(ii) $_4Be^9$ ***(p, α) ?***

[**Ans.** (*i*) $_3Li^7$(p, n) $_4Be^7$, Q = – 1.645 MeV.

(*ii*) $_4Be^9$ (p, α) $_3Li^6$, Q = 2.125 MeV.]

3

Nuclear Force

Structure of Nucleus

Nuclear Sizes and Shapes : The Rutherford α-scattering experiment established that the mass of an atom is concentrated within a small, positively-charged region at the centre which is called the 'nucleus' of the atom. Since Rutherford's time many scattering experiments, using highly energetic electrons and neutrons as the scattering particles, have been performed to determine the size of the nucleus. An electron interacts with a nucleus only through electric forces, while a neutron interacts only through nuclear forces. Thus electron scattering tells the distribution of charge in a nucleus and neutron scattering tells the distribution of nuclear mass. These experiments have shown that the volume of a nucleus is directly proportional to the number of nucleons in it, which is its mass number A.

If a nuclear radius is R, the volume is $\frac{4}{3}\pi R^3$ and so R^3 is proportional to A. This relationship is written as

$$R = R_0 A^{1/3}.$$

The value of the constant R_0 is experimentally found to be given by

$$R_0 \simeq 1.2 \times 10^{-15} \text{ m.}$$

The length 10^{-16} m is described as 1 femtometer (fm) or 1 fermi. Thus

$$1 \text{ fermi} = 10^{-15}\ m = 10^{-5}\ \text{Å}.$$

Hence we can write

$$R = 1.2\ A^{1\backslash 3} \text{ fm}$$

for nuclear radii. The radius varies from nucleus to nucleus and is of the order of 1 fm or 10^{-15} m.

While the size of the nuclei are readily determined by experiment, not too much is known about the shape of nuclei. For almost all purposes nuclei may be regarded as being 'spherical'. Certain nuclei, however, deviate from sphericity but the deviation is only 10 percent, or so.

Structure of the Nucleus—Proton-Electron Hypothesis : Before the discovery of the neutron, the nucleus of an atom was thought to be composed of protons and electrons which could account for the mass and charge of the nucleus. For example, the helium nucleus, whose mass is four times that of the proton but its charge is only + 2*e*, was thought as being composed of 4 protons and 2 electrons. This proton-electron hypothesis of the nucleus could easily account for the β–emission from the radioactive nuclei. It, however, failed because of a number of strong arguments against it :

(i) *Nuclear Size :* The radii of the nuclei are only of the order of 10^{-15} meter. The uncertainty principle demands that an electron confined to such a small region must have a kinetic energy of the order of 100 MeV. However, the electrons emitted during β–emission have energies of only 2 or 3 MeV. Hence they cannot be present within the nucleus.

The uncertainty principle applied to protons in a nucleus demand a kinetic energy only of the order of 1 MeV which is acceptable.

(ii) *Nuclear Spin* : A proton and an electron each has a spin of 1/2. Thus nuclei with an even number of protons *plus* electrons should have integral spins, while those with an odd number should have half-integral spins. This prediction is, however, not obeyed. The deuteron (nucleus of deuterium) has a mass about two times that of proton, but a charge of only + *e*. Therefore, on the proton-electron hypothesis, it must have two protons and one electron *i.e.* an odd number of particles. Its spin should therefore be half-integral $\frac{1}{2}$ or $\frac{3}{2}$ depending on the orientation of the particles). However, the observed spin of the deuteron is 1. Again, this shows that the electron as a separate particle cannot exist within the nucleus.

(iii) *Nuclear Magnetic Moment* : An electron, owing to its spin, has a magnetic moment of the order of a Bohr-magneton. Where there is an odd number of electrons, the odd electron should add this amount of magnetic moment to the nucleus. Yet the magnetic moment of nuclei are only about 1/1000 th of this. Hence the electrons cannot be the nuclear constituents.

In addition to the above theoretical considerations, experiments on electron scattering by nuclei and on nuclear disintegrations also rule out the existence of electrons in nuclei. Nevertheless, there remains the fact that many radioactive nuclei emit fast electrons (β–rays).

Proton-Neutron Hypothesis : In 1932 came the discovery of neutron, an *uncharged* particle of mass slightly more than that of a proton, and a spin of $\frac{1}{2}$. All these properties fit perfectly with the observed properties of nuclei when it is assumed that nuclei contains protons and neutrons tightly packed together (Fig.). The protons provide the (positive) charge, and the protons and neutrons together provide the mass of the nucleus. Thus the nucleus of hydrogen atom consists of 1 proton only. The nucleus of helium atom has 2 protons and 2 neutrons. In general, the nucleus of an atom of atomic number Z and mass number *A* contains Z protons and *(A–Z)* neutrons. The nuclear particles, proton and neutron, bear a common name 'nucleon'.

The hypothesis that nuclei are composed of protons and neutrons explains the existence of isotopes. The nuclei of all the isotopes of an element contain the same number of protons but have different number of neutrons. For example, ordinary hydrogen has three isotopes : hydrogen, heavy hydrogen (deuterium) and tritium. The hydrogen nucleus has only one proton, the heavy-hydrogen nucleus (called deuteron) has one proton and one neutron, while tritium nucleus (called tritron) has one proton and two neutrons.

The proton-neutron hypothesis also explains the observations regarding nuclear stability. Not all combinations of protons and neutrons form stable nuclei. There are three observations :

(i) Light nuclei contain roughly equal numbers of protons and neutrons. For heavier nuclei, the neutron number becomes progressively larger than the proton number. For the heaviest stable nuclei, the neutron number exceeds the proton number by about 50%. The excess of neutrons over protons in the heavier nuclei is necessary because protons repel one another electrostatically, and so their relative number should be less in order to secure the stability of the nucleus.

(ii) Nuclei of atoms beyond $_{83}Bi^{209}$ are unstable and undergo radioactive decay. This is so because when the proton number exceeds 83, the coulomb repulsion becomes so large that the neutrons are not able to prevent the disruption of the nucleus.

(iii) Sixty percent of the stable nuclei have even number of protons and even number of neutrons. (These are called even-even nuclei). Nearly all the others have either even number of protons and odd number of neutrons (even-odd nuclei) or vice versa (odd-even nuclei). Only five stable nuclei are odd-odd. Thus there is a preference for "even" numbers of protons or neutrons. This is in consistent with the existence of nuclear energy levels, each capable of containing two particles (neutrons or protons) of opposite spins.

Packing Fraction : The masses of the atomic isotopes are all very close to whole numbers, but are not *exactly* whole numbers. The difference between the actual atomic mass M of an isotope and its mass number A divided by the mass number is defined as the packing fraction P of the isotope. Thus

$$P = \frac{M - A}{A}$$

The packing fraction is positive for isotopes of very low and very high mass numbers, and negative for the rest. It is zero for $_6C^{12}$, by definition.

Mass Defect and Binding Energy : The masses of *all* stable nuclei are less than the sum of the masses of their constituent particles (protons and neutrons) in the free state. This means that when protons and neutrons combine to form a nucleus, a loss in mass results. The "missing" mass is released in the form of energy when the nucleus is formed. If Δm be the missing mass in the formation of a nucleus, the energy released ΔE will be given by the Einstein's relation :

$$\Delta E = (\Delta m)\, c^2$$

where *c* is the speed of light. The missing mass (Am) is called the 'mass defect', and its energy equivalent *(ΔE)* is called the 'binding energy' of the nucleus. Δ *E* is actually the energy that must be supplied to the nucleus to break it into its constituent particles.

The greater its binding energy, the more the energy that would be needed to break the nucleus. Thus, the binding energy is a measure of the stability of the nucleus.

Binding Energy of Deuteron : Deuteron is the simplest nucleus containing more than one nucleon. It consists of a proton and a neutron. The mass of the deuteron is less than the sum of the masses of proton and neutron. The energy equivalence of this mass difference is the binding energy of the deuteron. Then, if m_n , m_p and m_d be respectively the masses of neutron, proton and deuteron, the binding energy of the deuteron would be

$$E_B = (m_n + m_p - m_d)\, c^2. \qquad \dots (i)$$

In order to calculate the binding energy of the nucleus, we must use *nuclear* masses, *i.e.* neutron mass, proton mass and the mass of the nuclide.

We can write the nuclear binding energy in terms of atomic masses also. For example, the above equation will be written as

$$E_B = [m_n + m\,({}_1H^1) - m({}_1H^2)]\, c^2 \qquad \dots (ii)$$

where m (${}_1H^1$) stands for the mass of hydrogen *atom* and m (${}_1H^2$) for the mass of deuterium *atom*. The nuclear mass of hydrogen (the proton mass) is just the *atomic* mass of hydrogen minus the mass of one electron. The nuclear mass of deuterium (the deuteron mass) is the *atomic* mass of deuterium minus the mass of one electron. Thus electron mass automatically cencels out in calculating the nuclear binding energy from eq. (*ii*).

In general, the *nuclear* binding energy of an atom ${}_Zx^A$ can be written in terms of atomic masses in the following way :

$$E_B = [(A - Z)m_n + Zm({}_1\mathrm{H}^1) - m\,({}_Z\mathrm{X}^A)]\,c^2.$$

Substituting $m_n = 1.675 \times 10^{-27}$ kg, $m_p = 1.672 \times 10^{-27}$ kg,

$m_d = 3.343 \times 10^{-27}$ kg and c

$= 3.0 \times 10^8$ m/s in eq. (*i*), we get

$$E_B = (1.675 + 1.672\ 3.343) \times 10^{-27}\ \text{kg} \times (3.0 \times 10^8\ \text{m/s})^2$$

$$= 0.036 \times 10^{-11}\ \text{joule}$$

$$= \frac{0.036 \times 10^{-11}}{1.6 \times 10^{-13}} = 2.25\ \text{MeV}.$$

Thus the deuteron has a binding energy of 2.25 MeV. Using more accurate values of masses, the binding energy is about 2.23 MeV. This is confirmed by photo disintegration experiments which show that a γ-ray photon with a minimum energy of 2.23 MeV can split a deuteron into a free neutron and a free proton.

Binding Energy Curve : The nuclear binding energies range from 2.23 MeV for the heavy hydrogen to about 1800 MeV for Uranium. This shows that the binding energy increases with the complexity of the nucleus. Therefore, in order to compare the stabilities of different nuclei, we require the 'average binding energy per nucleon', which is obtained by dividing the total binding energy of a nucleus by the number of nucleons (protons+ neutrons) it contains. For example, the average binding energy per nucleon for, H^2 is

$$\frac{\text{total binding energy of } {}_1\mathrm{H}^2}{\text{mass number (no. of nucleons)}} = \frac{2.23\ \text{MeV}}{2} = 1.115\ \text{MeV}.$$

When the average binding energy per nuclear for various nuclei is plotted against the mass number *A*, a curve is obtained having the appearance as shown in Fig. This is known as the 'binding energy curve. The curve rises first rapidly and then slowly until it reaches a maximum of 8.8 MeV at $A = 56$, corresponding to the iron nucleus Fe^{56}. It then drops *very slowly* to about 7.6 MeV at ,4=238 (Uranium). Evidently *nuclei of intermediate mass are the most stable,* since they have the greatest average binding energy per nucleon (in the neighbourhood of 8.5 MeV). This means that the greatest amount of energy is needed to break them in their nucleons.

The (light) nuclei with $A < 20$ are the least stable. For example, the average binding energy per nucleon for the heavy hydrogen nucleus H^2 is only about 1 MeV. The subsidiary peaks occurring at He^4, C^{12}, O^{16} (even-even nuclei) indicate that these nuclei are much more stable than their immediate neighbours.

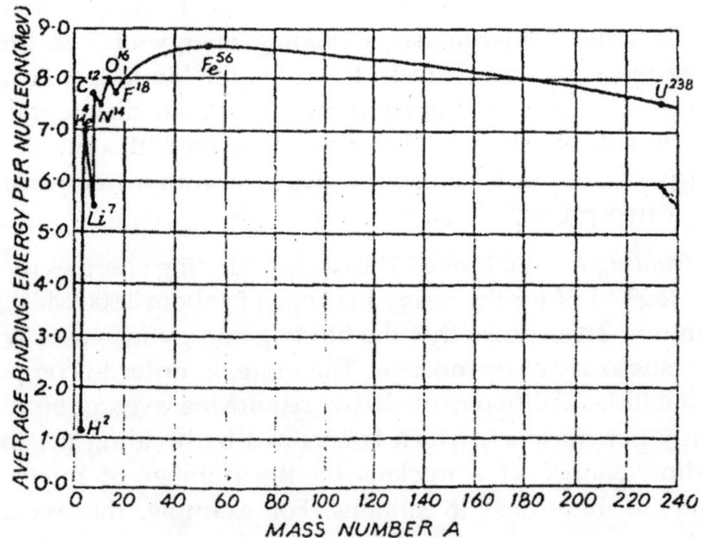

Explanation of Nuclear Fission and Nuclear Fusion : The general appearance of the binding energy versus mass number curve leads us to two remarkable conclusions which are of immense practical importance. The curve shows that very heavy and very light nuclei, which lie on either side of the broad maximum of the curve, have a lower average binding energy per nucleon. Therefore, if a very heavy nucleus (such as U^{238}) be splitted into two lighter nuclei near the top of the curve, the binding energy per nucleon will increase by about 1 MeV. This will obviously result in a general release of energy. This method of releasing nuclear energy by breaking up a heavy nucleus is known as 'nuclear fission', and is the basis of nuclear bombs and nuclear reactors.

Alternatively, if two or more very light nuclei (such as H^2) be combined into a heavier nucleus (say He^4), the binding energy per nucleon will again increase and now by a much greater amount than in the fission process. This will result in a much larger release of energy. This method of releasing nuclear energy is called 'nuclear fusion', and is the source of energy in the sun and in stars.

Power of Nucleus

The protons which repel each other electrostatically, and the neutrons which exert no electric force are held extremely close together inside a stable nucleus. Obviously, there are some forces operating within the nucleus which bind neutrons to protons (*n-p* forces), protons to protons (*p-p* forces), and neutrons to neutrons (*n-n* forces), thus holding the nucleus in one piece. Experimental evidences lead to the following properties and the nature of these nuclear forces :

(i) *Nuclear Forces are Primarily Attractive* : The overall effect of the nuclear forces is attractive, otherwise the nucleus would be disrupted under the electrostatic repulsion between the protons. (However, there is a repulsive component also in nuclear forces because an exclusively attractive force would lead to a collapse of the nucleus into a volume much smaller than the observed value).

(ii) *Nuclear Forces are Non-electric:* The nuclear forces cannot be of an electrical nature. If it were so, the protons would repel one another thus leading to nuclear disruption rather than to stability.

(iii) *Nuclear Forces are Non-gravitational :* If the gravitational forces between the nucleons are calculated, they are found to be about 10^{-40} times than the attractive forces demanded. Thus the nuclear forces cannot be gravitational in origin. In general, the nuclear forces are not central forces.

(iv) *Nuclear Forces are Extremely Strong :* The study of simplest stable nuclei, like $_1H^2$, shows that the forces holding the nucleons together must be very strong. In fact, they constitute by far the strongest class of forces known.

(v) *Nuclear Forces are Extremely Short-range Forces :* The nuclear forces do not obey the inverse-square law, and act only over distances as short as a small multiple of 10^{-15} meter. At distances appreciably smaller than 10^{-15} meter, the attraction of nucleons is replaced by repulsion.

(vi) *Nuclear Forces are Charge-independent :* The nuclear forces make no distinction between neutrons and protons. Experiments involving the scattering of protons and neutrons show that the (attractive) nuclear forces existing between protons and protons (*p-p* forces), between neutrons and neutrons (*n-n* forces), and between protons and neutrons (*p-n* forces) are all essentially the same in magnitude.

(vii) *Nuclear Forces are Spin-dependent :* Experimental evidences show that the nuclear force acting between two nucleons depends on the mutual orientation of the spins of the nucleons. For example, a neutron and a proton are kept together forming a deuteron only if their spins are parallel to each other.

(viii) *Nuclear Forces have the Property of Saturation :* This means that in the nucleus any one nucleon interacts with only a limited number of other nucleons nearest to it (*not* with all the other nucleons of the nucleus). This is apparent from the fact that the average binding energy per nucleon, after rising rapidly among the lightest nuclei, remains approximately constant for further increase in mass number A. This means that the total binding energy is proportional to the mass number A. When a fresh nucleon is added to the nucleus, it interacts with a *fixed* limited number of nearest nucleons, and so adds a *constant* amount to the binding energy.

Protons in the Nucleus do not Fly Apart : The very Strong attractive nuclear forces (p-p, n-n and n-p) between nucleons overcome the relatively weak electrostatic repulsive forces between protons. Hence the nucleus remains intact.

The electrostatic repulsion between protons make, however, the nucleus less stable than it otherwise would be. The emission of α-particles from heavier nuclei (which have a large number of protons) indicates this fact.

Yukawa's Meson Theory of Nuclear Forces : The mechanism responsible for the strongly-attractive, non-electrical and non-gravitational, short-range forces between nucleons remained a mystery for quite long. Heisenberg; in 1932, had suggested that the nuclear forces are 'exchange forces' which arise by the exchange of electrons (or positrons) between nucleons. But, calculations based on β–decay data snowed that the forces resulting from electron and positron exchange were 10^{14} times weaker than those actually required for nuclear stability.

Yukawa, in 1935, predicted the existence of a new particle, now called a pi-meson, having a rest mass greater than that of an electron but less than that of a nucleon. He proposed a theory of nuclear forces which involves these pi-mesons and is known as the 'meson theory of nuclear forces.' According to this theory, all nucleons (protons and neutrons) consist of identical cores surrounded by a 'cloud' of one or more pi-mesons. The mesons may be neutral, or may carry either a positive or a negative charge equal to the electronic charge, and are designated as π^o, π^+ and π^- respectively. The sole difference between protons and neutrons lies in the composition of their respective meson clouds. The mesons are being continually exchanged between nearby nucleons. The (nuclear) force between a neutron and a proton is the result of the exchange of charged mesons (π^+ and π^-) between them. When a π^+ meson jumps from a proton to a neutron, the proton is converted into a neutron, and the neutron is converted into a proton :

$$p - \pi^+ \rightarrow n$$

and $$p + \pi^+ \rightarrow p.$$

Conversely, when a π^- meson jumps from a neutron to a proton, the neutron is converted into a proton, and the proton is converted into a neutron :

$$n - \pi^- \rightarrow p$$

and $$p + \pi^- \rightarrow n.$$

In the same way, the forces between two protons and those between two neutrons arise by the exchange of neutral mesons (π^o) between them. Thus the nucleus is an ever-changing structure. However, the numbers of protons and neutrons remain the same in a particular structure.

A question arise at this point : When mesons have a finite rest mass, why this mass is not added to the mass of protons and neutrons ? The answer lies in the uncertainty principle. The mesons are constantly running between the nucleons. That is, the nucleons are constantly emitting and absorbing mesons. As such, even in principle, it is impossible to detect any change in mass of the nuclei's.

The actual discovery of pi-mesons in cosmic radiation, after Yukawa's prediction of its existence, opened a new field of research in nuclear structure.

Estimation of π-meson Mass : Let m_π be the mass of the π-meson. Its energy equivalence is $\Delta E = m_\pi\, c^2$. Suppose the meson travels between nucleons at a speed v (≈c). Nuclear forces have a maximum range r of about 1.7 fm (= 1.7×10^{-15} m). The time Δt required for the meson to travel this distance is

$$\Delta t = \frac{r}{v} \approx \frac{r}{c}$$

From the uncertainty relation, we have

$$\Delta E\ \Delta t \approx \frac{h}{2\pi}$$

or $$(m\pi c^2)\left(\frac{r}{c}\right) \approx \frac{h}{2\pi}$$

or $$m_\pi \approx \frac{h}{2\pi}\frac{1}{rc}.$$

Substituting the values, we get

$$m_\pi \approx \frac{6.63 \times 10^{-34}\ \text{j-s}}{2 \times 3.14 \times \left(1.7 \times 10^{-15}\ \text{m}\right) \times \left(3.0 \times 10^{8}\ \text{m/s}\right)}$$

$$\approx 2.1 \times 10^{-28}\ \text{kg}.$$

This is about 230 times the rest mass of the electron (9.1×10^{-31} kg).

Nuclear Models : An atomic nucleus shows a number of properties: It has (*i*) a radius proportional to $A^{1/3}$ (where A is the mass number), (*ii*) a well-defined and uniform density, (*iii*) a binding energy per nucleon which is almost constant (except for very small A) and (*iv*) a neutron to proton ratio which is close to unity for small A but increases progressively with increasing *A*. Further, it shows many types of nuclear reactions, radioactive decay and also the phenomena of nuclear fission and nuclear fusion. Many nuclear models have been suggested in an attempt to explain these properties although none of them is entirely satisfactory. Of these, two are more important : the liquid-drop model and the shell model.

Liquid-drop Model : Bohr, in 1936, pointed out that the strong, short-range, attractive forces between the nucleons are analogous to those acting between the molecules of a liquid. Therefore, *the nucleus is like a liquid-drop in which the nucleons are closely packed together, each nucleon being strongly attracted only by its immediate neighbours.* On this model, the nucleus will be a sphere with a radius proportional to $(\text{mass})^{1/3}$, and its density will be independent of its volume.

Let us now see how the liquid-drop model accounts for the observed variation of binding energy per nucleon with mass number. We know that the binding energy arises from the proton-proton, neutron-neutron, and proton-neutron nuclear attractions and is proportional to the number of nucleons *A* (the mass number). \Ve may write it as

$$Ev = a_1 A,$$

where a_1 is a positive constant. The energy E_v is called the volume energy of the nucleus.

Now, treating the nucleus as a liquid-drop, the binding energy is decreased by a surface term, because the nucleons on the surface have fewer neighbours to attract than those in the interior. This is proportional to the surface area, so that

$$E_s = -a_2 A^{2/3}.$$

The negative energy E_s is called the surface energy of the nucleus. It is most significant for the lighter nuclei since a greater fraction of their nucleons are on the surface.

The electrostatic repulsion between each pair of protons in the nucleus also contributes toward decreasing its binding energy. The coulomb energy is proportional to the number of proton pairs Z(Z—1)/2 ! in the nucleus, and inversely proportional to the nuclear radius ($\propto A^{1/3}$), so that

$$E_c = -a_3 \frac{Z(Z-1)}{A^{1/3}}.$$

The coulomb energy is negative because it arises from an effect that opposes nuclear stability.

The total binding energy of a nucleus is the sum of its volume, surface and coulomb energies :

$$E_B = E_V + E_S + E_c$$

$$= a_1 A - a_2 A^{2/3} - a_3 Z(Z-1) A^{-1}/3.$$

In a more complete equation two more energy terms are added *to* account for the excess of neutrons over protons in the heavier nuclei and for the occurrence of *proton pairs* and neutron pairs. These energies are called 'asymmetry energy' and 'pairing energy' respectively. The asymmetry energy is negative because it reduces the binding energy of the nucleus. The pairing energy is positive for even-even nuclei (which are most stable), zero for odd-even and even-odd nuclei, and negative for odd-odd nuclei. The average binding energy *per nucleon* is

$$\frac{E_B}{A} = a_1 - a_2 A^{-1/3} - a_3 Z(Z-1) A^{-4/3}.$$

In Fig., each of the terms on the right is plotted against A, together with their sum, E_B/A. The curve representing the *total* binding energy per nucleon is fairly close to the experimental curve. Hence the analogy of a nucleus with a liquid drop has some validity at least.

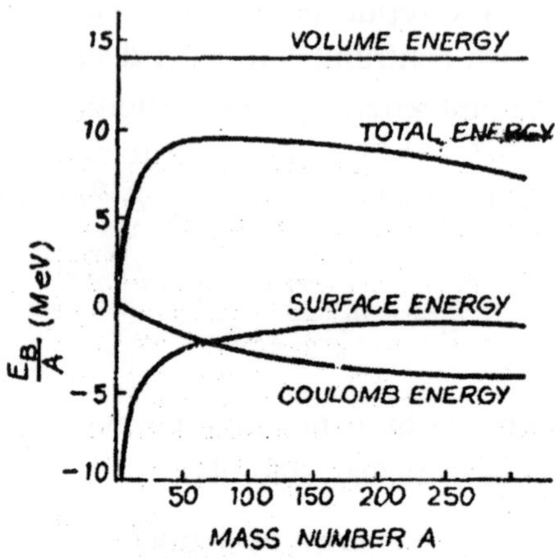

The nuclear reactions, accompanied by the emission of nucleons from the nucleus, can be interpreted on the basis of liquid-drop nuclear model. If the thermal agitation in the drop of a liquid is increased by heating, evaporation of molecules takes place. In a similar way, when an energetic particle is captured by a nucleus, a *compound* nucleus is formed which emits nucleons almost immediately. The emission of nucleons from a compound nucleus is thus analogous to the evaporation of molecules from a hot liquid drop.

Nuclear Fission and Liquid-drop Model : The liquid-drop model best explains the nuclear fission in which a heavy nucleus splits into two more or less equal parts. This occurs in the same way as a liquid-drop, if made to oscillate with sufficient energy, may become unstable and break.

The liquid-drop model fails, however, to explain nuclear reactions involving very high energies, and also the properties of the excited states of the nuclei.

Shell Model—Magic Numbers : There are certain nuclear phenomena which show the existence of excited states of nuclei and of sharply defined nuclear energy levels. This led to the shell model of the nucleus, the idea of which was borrowed from external electronic structure of the atom.

We know that the electrons in an atom are grouped in "shells", and "sub-shells", and atoms with 2, 10, 18, 36, 54, and 86 electrons have all their shells completely filled. Such atoms are unusually stable and are chemically inert. A similar situation exists with nuclei also. *Nuclei having 2, 8, 20, 28, 50, 82 and 126 nucleons of the same kind (either protons or neutrons) are more stable than nuclei of neighbouring mass numbers.* These numbers are called as 'magic numbers'. This observation is based on the following lines :

(i) Helium (Z = 2) and oxygen (Z=8) are particularly stable.

(ii) Calcium (Z = 20) and tin (Z = 50) have larger number of stable isotopes than any other element.

(iii) The end-products of the three naturally occurring radioactive series are lead isotopes (Z = 82).

(iv) The binding-energy curve has small peaks (showing extra stability) for the following nuclei :

$$_{28}Ni^{62},\ _{38}Sr^{88},\ _{50}Sn^{120},\ _{58}Ce^{140},\ _{82}Pb^{208}$$

It is seen that for these nuclei either Z (number of protons) or $A — Z$ (number of neutrons) is a magic number.

(v) The electric quadripole moments of magic number nuclei are very low (nearly zero) compared with those of other nuclei. This means that these nuclei have almost *spherical* charge distribution which is expected for the more stable nuclei.

The shell model of the nucleus is an attempt to account for the existence of magic numbers and certain other nuclear properties. According to this model, *the nucleons (protons and neutrons) in a nucleus live in a square potential well with rounded corners in which they occupy separate discrete sets of energy levels.* When a spin-orbit interaction is assumed to exist, the energy levels of either class of nucleon are arranged into different groups (shells). In this group-formation the gaps appear at just the right places, *i.e.,* after the completed groups having 2, 8, 20, 28, 50, 82, and 126 protons or neutrons.

The shell model is able to account for several nuclear phenomena in addition to magic numbers. For instance, it is observed that even-even nuclei are, in general, more stable than the odd-odd nuclei. This is obvious from the shell model. According to Pauli's principle, a single energy sub-level can have a maximum of two nucleons (one with spin up and other with spin down). Therefore, in an even-even nucleus only *completed* sub-levels are present which means greater liability. On the other hand, an odd-odd nucleus contains incomplete sub-levels for both kinds of nuclear which means lesser stability.

Further, the shell model is able to predict the total angular momenta of nuclei. Since the two protons (or neutrons) of each sub-level have *opposite* spins, they cancel each other's angular momentum. Hence, the actual total angular momentum of a nucleus depends only on the presence of an *extra* neutron or proton or both. In even-even nuclei, all protons and also all neutrons mutually cancel, so that the total angular momentum is zero for such nuclei. This has actually been observed with no exception.

The odd-odd nuclei have an extra proton and an extra neutron. Since each of these particles has an half-integral spin they give an integral total angular momentum, and this is experimentally observed. A similar reasoning predicts a half-integral total angular momentum for even-odd or odd-even nuclei, which is also experimentally confirmed.

Both the liquid-drop and shell models of the nucleus account, in their very different ways, different groups of nuclear phenomena. Neither is entirely satisfactory. Recently attempts have been made to combine the best features of each of these models and some success has been achieved. The resulting model is known as 'collective model'.

PROBLEMS

1. *Compute the approximate nuclear radius of Al^{27}.*

Solution: The nuclear radius of an atom of mass number A is given by

$$R = R_0 A^{1/3}$$

The proportionality constant R_0 is roughly 1.2 fermi. Thus

$$R = 1.2\ A^{1/3} \text{fm}.$$

For aluminium, $A = 27$.

$$\therefore \qquad R = 1.2\ (27)^{1/3}\ \text{fm}$$

$$= 3.6\ \text{fm} = 3.6 \times 10^{-15}\ \text{m}.$$

2. *A nucleus with A = 235 splits into two nuclei whose mass numbers are in the ratio 2:1. Find the radii of the nuclei. (Take R_0 = 1.4 fermi).*

Solution: The mass numbers of the resulting nuclei are 157 and 78.

The radius of a nucleus of mass number A is given by

$$R = R_0\ A^{1/3}$$

$$= (1.4\ \text{fm})\ A^{1/3}. \qquad [\text{Given}: R_0 = 1.4\ \text{fm}]$$

For $A = 57$, we have

$$R = (1.4\ \text{fm})\ (157)^{1/3} = 7.5\ \text{fm}.$$

For $A = 78$, we have

$$R = (1.4\ \text{fm})\ (78)^{1/3} = 6.0\ \text{fm}.$$

3. *What is the mass number A of a nucleus whose radius R is 2.5 fm ?*

Solution: $R = R_0\ A^{1/3}$,

where R_0 is roughly 1.2 fm (1 fm = 10^{-15} m). Here R = 2.5 fm.

$$\therefore \qquad 1.2\ A^{1/3} = 2.5$$

$$\therefore \qquad 1.2\ A^{1/3} = 2.5$$

$$\text{or} \qquad A^{1/3} = \frac{2.5}{1.2} = 2.08,$$

$$\therefore \qquad A = (2.08)^3 = 9.$$

4. ***Estimate the density of nuclear matter in kg/m^3, given that the nuclear radius R≈1.2 A$^{1/3}$ fm. Take mass of the nucleus A amu where A is mass number and 1 amu = 1.66 × 10^{-27} kg. (1 fm = 10^{-15} meter.)***

Solution: The mass of the nucleus is

$$M = A \text{ amu} = A \times 1.66 \times 10^{-27} \text{ kg}.$$

Treating the nucleus as sphere, its volume is

$$V = \frac{4}{3}\pi R^3 \approx \frac{4}{3} \times 3.14 \times (1.2\ A^{1/3} \times 10^{-15} \text{ meter})^3$$

$$\frac{4}{3} \times 3.14 \times (1.2)^3\ A \times 10^{-45} \text{ meter}^3$$

$$- 7.23 \times 10^{-45}\ A \text{ meter}^3.$$

Therefore, the density of nuclear matter is given by

$$\rho = \frac{M}{V} \approx \frac{A \times 1.66 \times 10^{-27} \text{ kg}}{7.23 \times 10^{-15} \text{ A meter}^3}$$

$$\approx 2.3 \times 10^{17} \text{ kg/m}^3.$$

5. ***Calculate the binding energy per nucleon in $_6C^{12}$. Masses of proton, neutron and electron are 1.00728, 1.00867 and 0.00055 amu respectively. The mass of $_6C^{12}$ atom is 12.00000 amu.***

Solution: The nucleus of $_6C^{12}$ atom contains 6 protons and 6 neutrons. Now,

mass of 6 protons = 6 ×1.00728 = 6.04368 amu

mass of 6 neutrons = 6 × 1.00867 = 6.05202 amu

total = 12.09570 amu.

This is the mass of the constituent particles of the carbon nucleus.

The carbon atom has 6 electrons. Now,

$$\text{mass of } {}_6C^{12} \textit{ atom} = 12.00000 \text{ amu}$$

$$\text{mass of 6 electrons} = 6 \times 0.00055 = 0.00330 \text{ amu}$$

$$\text{differences} = 11.99670 \text{ amu.}$$

This is the mass of the ${}_6C^{12}$. *nucleus*. Therefore, the mass defect is

$$\Delta m = \text{mass of nucleons—mass of nucleus}$$

$$= 12.09570 - 11.99670 = 0.09900 \text{ amu}$$

We know that 1 amu is equivalent to 931.5 MeV of energy. Therefore, the energy equivalent of Δm is

$$\Delta E = 0.09900 \times 931.5 = 92.2 \text{ MeV.}$$

This is the binding energy of carbon nucleus which has 12 nucleons. Therefore binding energy per nucleon is

$$\frac{92.2}{12} = 7.68 \text{ MeV.}$$

6. ***Taking the result of the last problem, calculate the binding energy of the least strongly bound neutron in the nucleus of ${}_6C^{13}$ isotope. The mass of the isotope is 13.00335 amu.***

*[**Ans.** 4.95 MeV]*

7. ***The atomic mass of ${}_6C^{12}$ is 12.000000 amu and that of ${}_6C^{13}$ is 13.003354 amu. Find the energy required to remove a neutron from ${}_6C^{13}$ in MeV. The mass of neutron is 1.008665 amu.***

Solution: The *nuclear* equation is

$${}_6C^{13} \rightarrow {}_0C^{12} + {}_0n^1$$

The mass of ${}_6C^{12} + {}_0n^1$ is 12.000000 + 1.008665 = 13.008665 amu.

$$\therefore \quad \text{mass defect} = 13.008665 - 13.003354$$
$$= 0.005311 \text{ amu.}$$

Its energy equivalent is

$$\Delta E = 0.005311 \times 931.5 = 4.95 \text{ MeV.}$$

This is the required energy.

8. ***Compute the binding energy of the "last" proton in a $_6C^{12}$ nucleus if the mass of this nucleus is 11.99671 amu and the mass of the $_5B^{11}$ nucleus is 11.00657 amu. The mass of proton is 1.00728 amu.***

Solution: The *nuclear* equation is

$$_5B^{11} + {}_1p^1 \rightarrow {}_6C^{12}.$$
(proton)

The mass of $_5B^{11} + {}_1P^1$ is 11.00657 + 1.00728 = 12.01385 amu, while that of the product nucleus $_6C^{12}$ is 11.99671 amu. Therefore, mass loss is

$$\Delta m = 12.01385 - 11.99671 = 0.01714 \text{ amu.}$$

Its energy equivalent is

$$\Delta E = 0.01714 \times 931.5 = 16 \text{ MeV.}$$

This is the binding energy of the last proton.

9. ***Calculate the binding energy of α-particle in MeV. The masses of the proton, neutron and the α-particle (helium nucleus) are 1.00728,1.00867 and 4.00151 amu respectively.***

Solution: The α-particle ($_2He^4$) is composed of two protons and two neutrons. The sum of the masses of its constituent particles is as follows :

$$\text{mass of 2 protons} = 2 \times 1.00728 = 2.01456 \text{ amu}$$
$$\text{mass of 2 neutrons} = 2 \times 1.00867 = 2.01734 \text{ amu}$$
$$\text{total} = 4.03190 \text{ amu.}$$

Subtracting from this sum, the mass of the α-particle (He nucleus), we obtain the mass-defect Δm, that is,

$$\Delta m = 4.03190 - 4.00151 = 0.03039 \text{ amu.}$$

1 amu is equivalent to 931.5 M.eV of energy. Therefore the energy equivalent of the above mass defect is

$$\Delta E = 0.03039 \times 931.5 = 28.3 \text{ MeV.}$$

This is the energy that would be liberated if one were to make an α-particle out of protons and neutrons. Conversely, this is also the energy that would have to be supplied to an α-particle to break it into protons and neutrons, and consequently it is the binding energy of the α-particle. This *large* amount of energy explains why the α-particle is a highly stable particle.

The α-particle has 4 nucleons. Hence its binding energy per nucleon is $\frac{28.3}{4} = 7.07$ MeV.

10. ***How much energy must a photon have if it is to split an α-particle ($_2He^4$ nucleus) into a triton ($_1H^3$ nucleus) and a proton ($_1H^1$ nucleus) ? The masses of the α-particle, triton and proton are 4.00151, 3.01550 and 1.00728 amu respectively and 1 amu is 931.5 MeV.***

Solution: The energy required for the photon is the binding energy of the α-particle formed by the fusion of triton and proton.

The mass of triton *plus* proton is 3.01550 + 1.00728 = 4.02278 amu, while the mass of the resulting α-particle is 4.00151 amu. Thus, the missing mass is

$$\Delta m = 4.02278 - 4.00151 = 0.02127 \text{ amu.}$$

Its energy equivalent is

$$\Delta E = 0.02127 \times 931.5 = 19.8 \text{ MeV.}$$

This is the binding energy, and hence the required energy for the splitting of the α-particle.

11. ***Calculate the binding energy of a deuteron (nucleus of $_1H^2$) from the following data :***

Mass of deuterium atom $_1H^2$ = 2.014102 u, mass of hydrogen atom $_1H^1$ =1.007825 u ana mass of neutron = 1.008665 u. (l u= 1.66 × 10^{-27} kg.)

Or

Calculate the minimum energy of gamma-rays necessary to disintegrate a deuteron into a proton and a neutron.

Solution: The deuterium atom $_1H^2$contains one proton and one neutron in its nucleus (called as deuteron) and one extra-nuclear electron. Let us calculate the mass defect of deuteron. We have

mass of $_1H^1$ atom (1 proton and 1 electron) = 1.007825 u

mass of 1 neutron = 1.008665 u

total = 2.016490 u.

This is the mass of all the constituent particles of the atom $_1H^2$. The mass of the atom $_1H^2$ is 2.014102 u. Therefore, the mass defect of its *nucleus* is

$$\Delta m = 2.016490 – 2.014102 = 0.002388 \text{ u.}$$

Now, $$1 \text{ u} = 1.66 \times 10^{-27} \text{ kg.}$$

$\therefore$ $$\Delta m = 0.002388 \times (1.66 \times 10^{-27})$$

$$= 3.964 \times 10^{-30} \text{ kg.}$$

Its energy equivalent is

$$\Delta E = (Am)\ c^2$$

$$= (3.964 \times 10^{-30} \text{ kg}) \times (3.0 \times 10^8 \text{ m/s})^2$$

$$= 3.567 \times 10^{-13} \text{ joule.}$$

Now, $$1 \text{ MeV} = 1.6 \times 10^{-13} \text{ joule.}$$

$\therefore$ $$\Delta E = \frac{3.567 \times 10^{-13}}{1.6 \times 10^{-13}} = 2.23 \text{ Mev.}$$

This is the (nuclear) binding energy of the deuteron.

When a deuteron is formed from a free proton and a free neutron, then 2.23 MeV of energy is liberated. Conversely, a gamma-ray photon must have an energy of at least 2.23 MeV to break a deuteron into a proton and a neutron.

12. ***Calculate the average binding energy per nucleon for a U^{238} nucleus in MeV. The mass of U^{238} atom is 238.05078 amu and it is composed of 92 protons and 146 neutrons. The masses of proton, neutron and electron are 100728, 1.00867 and 0.00055 amu respectively.***

Solution: Let us consider U^{238} atom. Its nucleus contains 92 protons and 146 neutrons. The sum of the masses of the constituent nucleons is as follows :

$$\text{mass of 92 protons} = 92 \times 1.00728 = 92.66976 \text{ amu}$$

$$\text{mass of 146 neutrons} = 146 \times 1.00867 = 147.26582 \text{ amu}$$

$$\text{total} = 239.93558 \text{ amu.}$$

The mass of the U^{238} atom is 238.05078 amu. It contains 92 electrons. Therefore, the mass of U^{238} nucleus is

$$238.05078 - (92 \times 0.00055) = 238.00018 \text{ amu.}$$

The mass defect is

$$\Delta m = 239.93558 - 238.00018$$

$$= 1.93540 \text{ amu.}$$

Its energy equivalent, that is, the binding energy of U^{238} nucleus is

$$\Delta E = 1.93540 \times 931.5$$

$$[\because 1 \text{ amu} = 931.5 \text{ MeV}]$$

$$= 1803 \text{ MeV.}$$

The total number of nucleons is 238. Therefore binding energy per nuclear is

$$\frac{1803}{238} = 7.57 \text{ MeV/nucleon.}$$

13. ***Find the total energy liberated in the reaction*** $_3Li^7$ ***(P, α) 2***H_e^4 ***if the binding energy per nucleon for*** Li^7 ***and*** He^4 ***are 5.61 MeV and 7.06 MeV respectively.***

Solution: The reaction is

$$_3Li^7 + {_1H^1} \rightarrow {_2He^4} + {_2He^4}.$$

The binding energy of $_3Li^7 = 5.61 \times 7 = 39.27$ MeV and that of $_2He^4 = 7.06 \times 4 = 28.24$ MeV. There is *no* binding energy in the nucleus of $_1H^1$ (it is simply a proton).

Thus, the total binding energy of the reactants is 39.27 MeV and that of the products is $2 \times 28.24 = 56.48$ MeV. The increase in binding energy is

$$56.48 - 39.27 = 17.21 \text{ MeV}.$$

This is the total energy liberated.

14. ***The binding energy of*** $_{17}Cl^{35}$ ***nucleus is 298 MeV. Find its atomic mass. The mass of hydrogen atom (***H^1***) is 1.007825 amu and that of a neutron is 1.008665 amu. Given : 1 amu = 931.5 MeV.***

Solution: The $_{17}Cl^{35}$ atom has 17 protons and 18 neutrons in its nucleus, and 17 extra nuclear electrons. Its mass would be the sum of the masses of the constituent particles *minus* the mass equivalent of the binding energy of its nucleus. Now,

mass of 17 $_1H^1$ atoms *i.e.* 17 protons and 17 electrons

$$= 17 \times 1.007825 = 17.133025 \text{ amu}$$

$$\text{mass of 18 neutrons} = 18 \times 1.008665 = 18.155970 \text{ amu}$$

$$\text{total} = 35.288995 \text{ amu}.$$

The mass equivalent of the binding energy of $_{17}Cl^{35}$ nucleus is

$$\Delta m = \frac{298}{931.5}\ 0.319914 \text{ amu}.$$

$\therefore$ atomic mass of $_{17}Cl^{35}$

$$= 35.288995 - 0.319914 = 34.96908 \text{ amu}.$$

4

Molecular Physics

Atomic Physics

Orbital Magnetic Dipole Moment

Bohr Magneton : An electron revolving about a nucleus is a minute current loop and has a magnetic field. It thus behaves like a magnetic dipole.

Magnetic dipole moment,

$$\left|\vec{\mu}_l\right| = \frac{eva}{2c}\text{emu}$$

Orbital angular momentum, $\left|\vec{I}\right|$

$$\left|\vec{I}\right| = mva$$

Gyromagnetic ratio

$$= \frac{\left|\vec{\mu}_l\right|}{\left|\vec{I}\right|} = \frac{e}{2mc}$$

Bohr Magneton is, $\mu_B = \dfrac{eh}{4\pi m}$

Larmor Precession : When the atom is placed in an external magnetic field, the electron orbit processes about the field direction as axis. This precession is called the 'Larmor precession', and the frequency of this precession is called the 'Larmor frequency'.

Spectroscopic Terms and their Notations

- The quantised energy states of an atomic electron are described in terms of the quantum numbers n, l, s and j.
- The electron having orbital quantum number l = 0, 1, 2, 3, 4, ... are named as s, p, d, f, g.
- For example $n = 2, l = 0$ is in a 2s state

  $n = 3, l = 1$ is in a $3p$ state
- By spin orbit interaction, each energy level of a given l is splitted into two sub-levels corresponding to

$$j = l + s = l + \frac{1}{2}$$

$$j = l - s = l - \frac{1}{2}$$

- The number of different possible orientations of $\vec{l}$ and $\vec{s}$ and hence the number of different possible values of j is known as the 'multiplicity' of the term.

Stern-Gerlach Experiment : This experiment showed that the atoms passing through the field were oriented in space in discrete direction.

The transverse deflection suffered by the atoms is given by:

$$z = g_j m_j \frac{\frac{\partial B_z}{\partial z} \mu_B d^2}{6kT}$$

where $\frac{\partial B_z}{\partial z}$ → rate of change of field in the z-direction.

This experiment is also an evidence for the existence of

electron spin and a value $\frac{1}{2}$ is assigned to the spin quantum number $\left(s = \frac{1}{2}\right)$

Spin-Orbit Interaction : The spin-orbit interaction is an interaction between an electron's spin magnetic moment and the internal magnetic field of the atom which arises from the orbital motion of the electron through the nuclear electric field.

$$\Delta T = R_\infty \alpha^2 z^4 / n^3 l(l+1)$$

$$\alpha \rightarrow \text{Fine structure constant}$$

Quantum-Mechanical Relativity Correction : Besides the energy shifts due to spin orbit interaction, the relativistic effect is equally important as it produces energy shifts in the hydrogen atom comparable to those produced by spin-orbit interaction.

The relativistic term shifts is

$$\Delta T_r = -\frac{\Delta E_r}{hc}$$

$$= \frac{R_\infty \alpha^2 z^4}{n^3}\left(\frac{1}{l+\frac{1}{2}} - \frac{3}{4n}\right) cm^{-1}$$

Formation of Hydrogen

Lamb-Shift

- The discrepancy in the separation between the main components of the H_α line could be explained by assuming the state $2^2\ S_{1/2}$ to be about 0.03 cm^{-1} higher than $2^2\ P_{1/2}$ in contradiction to Dirac's theory which had shown them to have exactly the same energy.
- For hydrogen-like atoms the states of a particular *n* value having terms with the same j value but different *l* values, such as $2^2\ P_{1/2}$, and $2^2\ S_{1/2}$, are not degenerate, but are separated. Such a separation is known as 'Lamb-Shift'.

Formulation of Pauli's Principle

- According to Pauli "no two electrons in an atom can exist in the same quantum state". This is known as Pauli's exclusion principle.

  If the particles of the system are Bose particles, the proper wave function is—

$$\psi_{\text{Bose}}(1, 2) = \frac{1}{\sqrt{2}}\left[\psi_a(1)\,\psi_b(2) + \psi_b(1)\,\psi_a(2)\right]$$

(Symmetric)

- If the particles are Fermi particles, the wave function is

$$\psi_{\text{Fermi}}(1, 2) = \frac{1}{\sqrt{2}}\left[\psi_a(1)\,\psi_b(2) - \psi_b(1)\,\psi_a(2)\right]$$

(antisymmetric)

- If the particles are in same quantum state, a = b,

  then $\psi_{\text{Bose}}(1, 2) \neq 0$

  but $\psi_{\text{Fermi}}(1, 2) = 0$

  Two Bose particles can exist in the same quantum state, two Fermi particles cannot.

Result of the Hartree Theory

1. The electron for the average value of r (radius of shell) holding for a multi-electron atom would be

$$\bar{r} = \frac{n^2 a_0}{z_n}\left[1+\frac{1}{2}\left\{1-\frac{l(l+1)}{n^2}\right\}\right] \simeq \frac{n^2 a_0}{z_n}$$

   In multi-electron atoms the radii of the inner shells of small n are very small, e.g., the radius of the n = 1 shell of argon ($z_n \simeq z - 2 \simeq 16$) is very smaller than that of the n = 1 shell of hydrogen by approximately a factor of $\frac{1}{16}$.

2. For the outermost populated shell of an atom, Hartree calculations show that

$$z_n \simeq n$$

and so the corresponding radius is

$$\bar{r} \simeq \frac{n^2 a_0}{z_n} \simeq \frac{n^2 a_0}{n} \simeq n a_0$$

Thus the radius of the outermost shell of a multi-electron atom is roughly n times, the radius of the smallest orbit of hydrogen.

3. Like radius, we can approximately express the total energy of an electron in a multi-electron atom by the energy equation of a one-electron atom, where z is replaced by z_n, *i.e.*

$$E \simeq -\frac{2\pi^2 \mu z_n{}^2 e^4}{n^2 h^2}$$

4. For the outermost populated shell of an atom, we have $z_n \simeq z$ and corresponding energy is

$$E \simeq -\frac{2\pi^2 \mu e^4}{h^2}$$

Thus, the total energy of an electron in the outermost shell of any atom is comparable to that of an electron in the ground state of hydrogen atom.

5. The energy equation for a multi-electron atom contains $z_n{}^2$ in the numerator and n^2 in the denominator. Meaning: for a given atom the total energy E becomes less negative with increasing n.

6. In a one-electron atom the total energy E depends only on the quantum number n, while in a multi-electron atom it depends on l as well as on n.

Spectroscopic Terms : L-S and j-j Couplings

Terminology

State : The 'state' of an atom is the condition of motion of all the electrons specified by four quantum numbers.

Degenerate : If several states have the same energy.

Ground State : The state with lowest energy is the ground state.

Energy Level : A collection of states having the same energy in the absence of external magnetic or electric field constitutes an "energy level". The level with the lowest energy is the ground level.

Sub Level : An external field splits an energy level into several "sublevels" each characterised by one or more magnetic quantum number.

Term : A collection of levels characterised by an orbital angular momentum and multiplicity comprises a 'Spectroscopic term'.

e.g. : a^3D term means the weighted average energy of the $^3D_3$, $^3D_2$, and $^3D_1$ levels.

Configuration : The specification of the quantum numbers *n* and *l* for the orbitals of all the electrons of an atom is called the 'electron configuration' of the atom. *e.g.* the electron configuration of the ^6C atom : $1s^2\ 2s^2\ 2p^2$.

Equivalent Orbitals : Orbitals with the same *n* and *l* are 'equivalent'. The electrons in equivalent orbitals are called 'equivalent electrons'.

Statistical Weight : The number of distinct states in a specified collection is the 'statistical weight'. The statistical weight of a level is 2J + 1 ; for term it is (2S + 1) (2L + 1) ; for a single electron it is $2n^2$. Spectral lines are categorised according to the following nomenclature.

Component : A transition between two sublevels is called a 'Component'.

Line : A transition between two levels is a 'line'. A line is thus a blend of components.

Multiplet : A collection of transitions between two term is called a multiplet. A multiplet thus consists of a number of lines.

Resonance Line : Among the lines arising from transition between the ground level and higher levels, the line of lowest frequency is called the "resonance line".

Selection Rules for Multi-electron Atoms in L-S Coupling: In the electric-dipole transitions in multi-electron atoms, the selection rules are—

1. $\Delta l = \pm 1$ (for one electron)

   The parity of the configuration must change in an electric dipole transition (odd ↔ even).

   For two electron transition we would have

   $$\Delta l_1 = \pm 1 \; ; \; \Delta l_2 = 0, \pm 2$$

2. There is no restriction on the total quantum number n of either electron.

3. For the atom as a whole, the quantum numbers L, S and J must change as follows:

   $\Delta L = 0, \pm 1$ (In one-electron atoms $\Delta L = 0$ is not allowed)

   $$\Delta S = 0$$

   $$\Delta J = 0, \pm 1 \text{ but } J = 0 \not\leftrightarrow J = 0.$$

j - j Coupling

(a) As a result of the stronger spin-orbit interaction, the orbital and spin angular momentum vectors of each individual electron are strongly coupled together to form a resultant angular momentum vector $\vec{j}$ of magnitude $\sqrt{j(j+1)}\frac{h}{2\pi'}$ where $j = l-\frac{1}{2}$ and $l+\frac{1}{2}$, *i.e.*, j takes half integral values only.

(b) As a result of the residual electrostatic interaction and spin-spin corelation, the resultant angular momentum vector $\vec{j}$ of the individual electrons are less strongly coupled with one another to form the total angular momentum vector $\vec{j}$ of the atom, of magnitude $\sqrt{j(j+1)}\frac{h}{2\pi'}$ where the quantum number j has the values.

$$J = |\vec{j_1} + \vec{j_2} + \dots|_{min}, |\vec{j_1} + \vec{j_2} + \dots|_{min} + 1, \dots \ (j1 + j2 + \dots)$$

Selection Rules in j - j Coupling

1. If two electron jump then

$$\Delta l_1 = \pm 1, \Delta l_2 = 0, \pm 2$$

2. $\Delta j = 0, \pm 1$ for the jumping electron and $\Delta j = 0$ for all the other electrons.

3. For the atom as a whole

$$\Delta J = 0, \pm 1 \text{ but } J = 0 \not\leftrightarrow J = 0$$

The selection rules $\Delta S = 0$ and $\Delta L = 0, \pm 1$ no longer hold, since L and S are no longer good quantum numbers.

The Zeeman Effect

- When an atom is placed in an external magnetic field, the spectral lines it emits are split into several polarised components. This effect of magnetic field on the atomic spectral lines is called Zeeman effect.
- A singlet spectral line viewed normal to the field is split into three plane-polarised components ; a central unshift

line with the electric vector vibrating parallel to the field (called π component) and two other lines equally displaced one on either side with electric vector perpendicular to the field (called σ components). This is called a "Normal triplet" and effect is 'Normal Zeeman effect'.

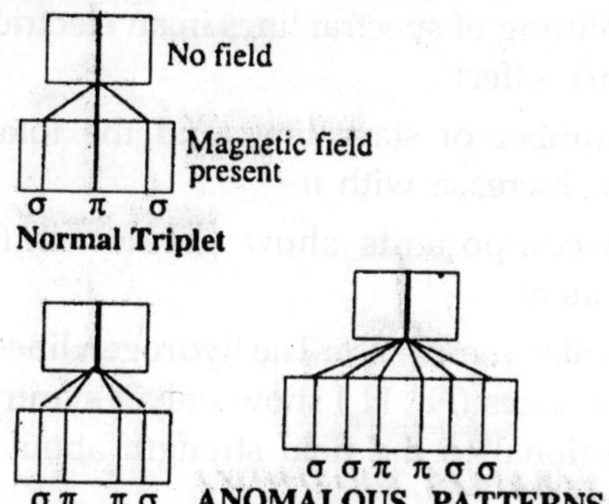

- The fine-structure components of a multiplet spectral line, however, show a complex Zeeman pattern, e.g., D_1 and D_2 components of sodium yellow doublet give four and six lines respectively in the Zeeman pattern. This is 'anomalous' Zeeman effect.

The Landed-factor is given by

$$g = 1+\frac{J(J+1)+S(S+1)-L(L+1)}{2J(J+1)}$$

Paschen-Back Effect : The strength of the external field is increased, the separations between zeeman components increase until they become greater than the separations between multiplet fine-structure components. The anomalous Zeeman pattern then changes over to like a normal Zeeman pattern. This phenomenon is known as 'Paschen-Back effect'.

In a strong field, the selection rules for transitions are

$$\Delta M_L = 0$$

(Components polarised parallel to the field)

$$\Delta M_L = \pm 1$$

(Components polarised perpendicular to the field)

$$\Delta M_S = +1$$

Stark Effect

- The splitting of spectral lines in an electric field is known as 'Stark effect'.
- The number of stark lines and the total width of the pattern increase with *n*.
- The π-components show greater shifts than the σ component.
- *First Order Stark Effect* : The hydrogen lines with the lower energy states (H_α, H_β) show only a symmetrical splitting proportional to the field strength about their field free position.
- *Second Order Stark Effect* : For the lines involving the higher states (H_γ, H_δ), the Stark components show unidirectional displacements proportional to the square of the field strength.
- The Stark-Shift for the ground state ($n = 1$) is zero.
- For a non-degenerate state, such as the ground state ($n = 1, l = 0, n_1 = 0$) of hydrogen atom which has even parity, there is no first order Stark effect.

  Only degenerate terms can have a first order Stark effect.

Nuclear Spin and Hyperfine Splitting

- Bismuth exits as a single isotope, but shows six hyperfine components in its line at wevelength 4722Å.

  The nuclear angular momentum is

$$|\vec{I}| = \sqrt{I(I+1)}\frac{h}{2\pi}$$

I → Nuclear spin quantum number.

Nuclear magnetic moment.

$$\mu_1 = g_I \sqrt{I(I+1)}\mu_N$$

$$\left[\mu_N = \frac{eh}{4\pi m_p c}\right]$$

- We have only to count the number of hyperfine components for any Zeeman line in a strong magnetic field, and this number is 2I + 1. Hence I is directly determined. Back and Goudsmit found that in a strong magnetic field each of the Zeeman components of Bi consists of 10 hyperfine components due to nuclear spin.

  Hence they concluded that for B_i, $I = \frac{9}{2}$.

X-Rays

- X-rays are produced when high speed electrons strike a metal of high atomic number.

X-rays spectrum consists of two spectra :

1. *Continuous Spectrum* : It is due to continuous deacceleration of the electron as it passes through the field of the nucleus. Its short-wevelength limit is given by—

$$\lambda_{min} = 12{,}400/V\text{Å}$$

2. *Line or Characteristic Spectrum* : It is produced when electrons of the target material are dislodged from their orbits by the striking electrons. The spectrum consists K-, L- and M-series etc. depending on when electron jump ends.

Moseley's Law : Moseley's law states that the frequency of a spectral line in the characteristic x-ray spectrum varies directly as the square of the atomic number of the element emitting it.

- Absorption of X-rays follows the exponential law

$$I = I_0 e^{-\mu x}$$

where μ is the absorption coefficient.

- The linear absorption coefficient μ may be defined as the fractional decrease in the intensity of the X-rays per unit thickness of the absorber material.
- Due to Compton scattering, change in wavelength of the scattered X-rays is given by—

$$\Delta\lambda = 0.024\ (1 - \cos\theta)\ \text{Å}$$

Spectrum of Molecules

The Born-Oppenheimer Approximation

- We cannot classify molecular energy state according to the electronic angular momentum L which is not conserved in molecules.
- The energy of each of the vibrational and rotational motions is quantised with the result that there are many more energy levels in a molecule than in an atom.
- We can solve the complete molecular wave equation in the following way.

  1. The electronic wave equation $H_e \psi_e = E_e' \psi_e$ for a given electronic state of the molecule is solved for a range of values of nuclear coordinates. This gives ψ_e and E_e as continuous functions of nuclear co-ordinates.
  2. Having obtained E'_e the nuclear wave equation $(H_n + E_e')\ \psi_n = E\psi_n$ is solved. The eigen values E are the characteristic energies for the whole molecule in the given electronic state.

Types of Molecular Spectra

Electronic Spectra : These spectra are observed, both in emisson and absorption in the visible and ultraviolet regions. The electronic spectra are observed for the heteronuclear as well as for the homonuclear diatomic molecules.

Vibrational-rotational Spectra : These spectra are observed in absorption, in the near infrared region ($1\mu - 10^2\ \mu$). They are observed only for the heteronuclear molecules. The homonuclear molecules like H_2, N_2, O_2, do not produce vibrational-rotational band.

Pure Rotational Spectra : These spectra are observed in absorption in the far infrared ($10^2\mu - 10^3\mu$) and observed only for the heteronuclear diatomic molecules.

Pure Rotational Spectra : Only molecules that have permanent electric dipole moments can absorb or emit electro-magnetic radiation in such transition.

Hamonuclear diatomic molecules such as H_2, O_2, N_2 ; symmetric linear molecules such as CO_2 and spherical top polyatomic molecules such as CH_4 do not exhibit rotational spectra.

- The moment of inertia of the molecule as a rigid rotator

$$I = \mu r^2$$

$\mu \rightarrow$ Reduced mass ;

$r \rightarrow$ distance from the axis of rotation.

- Eigen values, E are

$$E = \frac{h^2}{8\pi^2 I} J (J + 1)$$

J $\rightarrow$ Rotational quantum number;

$$J = 0, 1, 2, 3, ...$$

- In terms of wave number, the energy equation can be written as

$$\frac{E}{hc} = F(J)$$

$$= BJ(J + 1)cm^{-1} \text{ with}$$

$$B = \frac{h}{8\pi^2 Ic}$$

- Wave number

$$v = 2B(J + 1)$$

with $J = 0, 1, 2, ...$

$$v = 2B, 4B, 6B, ...$$

Thus, the absorption spectrum of a rigid rotator is expected to consists of a series of equidistant lines with constant separation 2B.

- The population in the Jth level relative to the lowest level;

$$\frac{N_J}{N_0} = (2J + 1)\ e^{-(E_J - E_0)/kT}$$

$$= (2J + 1)\ e^{-BJ(J+1)hc/kT}$$

Vibrational Rotational Spectra : In the series of lines which are not equidistant, a line is missing in the centre of the band. The missing line is known as the 'null line' or 'zero gap'.

The Molecule as a Harmonic Oscillator : The allowed energies for the harmonic oscillator $E = hv_{osc}\left(v + \frac{1}{2}\right)$

where $v = 0, 1, 2, ...$

- The vibrational terms

$$G(v) = \frac{E}{hc} = \frac{v_{osc}}{c}\left(v + \frac{1}{2}\right) = \omega\left(v + \frac{1}{2}\right)$$

where $\omega = \frac{v_{osc}}{c} \rightarrow$ Vibrational constant.

- Vibrational transitions can only occur when the molecule has a permanent dipole moment which changes with distance.
- The spacing between the vibrational levels are considerably larger than the spacings between the rotational levels of a molecule, intact larger than kT at room temperature.
- Thus, the main vibrational transition in absorption is

$$v = 1 \leftarrow v = 0$$

- The vibrational spectrum is expected to consists of a single band at ω cm^{-1}.
- In the vibrational spectrum, the nuclei carrying out harmonic vibrations along the internuclear axis.

Molecule as Anharmonic Oscillator

- The actual infrared spectrum is found to consists at an intense (fundamental) band at ω plus a number of weak bands (overtones) at slightly lesser and lesser than 2ω, 3ω, ... The observation of overtones indicates that the selection rule $\Delta v = \pm 1$ is not strictly obeyed and transitions corresponding to $\Delta v > 1$ do take place.
- The observation that the overtones appear not exactly at 2ω, 3ω, ... but at lesser and lesser values indicates that the vibrational energy levels are not exactly equally-spaced but converges slowly.
- The corresponding term values are given by

$$G(v) = \omega_e\left(v+\frac{1}{2}\right)-\omega_e x_e\left(v+\frac{1}{2}\right)^2 + ...$$

$\omega_e x_e \rightarrow$ the anharmonicity constant smaller than ω_e and always positive.

The vibrational frequency of anharmonic oscillator is

$$\therefore \quad v_{osc}(v) = c\,[(\omega_e - \omega_e x_e) - 2\omega_e x_e\, v]$$

The frequency

$$V_{osc}\left(-\frac{1}{2}\right) = c\omega_e \left[v = -\frac{1}{2} \text{ is hypothetical state}\right]$$

$\omega_e \rightarrow$ The vibrational frequency that anharmonic oscillator would have classically for an infinitesimal amplitude *i.e.*, in imaginary state $v = -\frac{1}{2}$ at the very bottom of the potential curve.

$$v_{osc} = c\omega_e = \frac{1}{2\pi}\sqrt{\frac{k_e}{\mu}}$$

or

$$k_e = 4\pi^2 \mu c^2\, \omega_e^2$$

Isotop Effect on Vibrational Level : The vibrational levels of the heavier isotope are lower than the corresponding levels of the lighter isotope.

$$G'(v) < G(v)$$

Molecule as Vibrating Rotator

Fine Structure of Infrared Bands : The near infrared spectra of molecules consists of 'band' (not lines) each band being composed of close lines arranged in a particular manner. This observed fine structure suggests that in a vibrational transition the molecule also changes its rotational energy state, *i.e.*, it must be treated as a vibrating-rotator.

- B, determines the spacing between the rotational levels, is different for different vibrational states.
- For a given vibrational transition, the rotational transitions $\Delta J = +1$ given one set of lines called the 'R-branch' while the rotational transitions $\Delta J = -1$ give the other set of lines called the 'P-branch'.

- $V_0 = G(v') - G(v'')$, is the wave number of the pure vibrational transition (for which $J' = J'' = 0$) which is unallowed ($\Delta J \neq 0$) and corresponds to the missing line in the band. V_0 is known as the wave number of the 'band-origin'.

- If we neglect vibration-rotation interaction, then

$$V_R = V_0 + 2B + 2BJ''$$

[Wave number of R branch]

$$V_P = V_0 - 2BJ''$$

[Wave number of P branch]

- The only case of a diatomic molecule for which a Q-branch (*i.e.* a central 'line') has been observed in the rotation-vibration infrared bands is that of nitric oxide NO.

Electronic Spectra : Franck-Condon Principle

- Each electronic states of the molecule has a different potential energy curve.
- A single electronic transition in a molecule gives rise to a band system.

Electronic Band Spectra in Absorption : The study of absorption spectra, as compared to emission spectra, gives more information, with less elaborate computation, regarding the excited electronic states of molecules.

- The absorption spectrum of I; molecule, show progressions corresponding to $v'' = 1$ and even higher, in addition to the progression $v'' = 0$, although with smaller intensity.

- The P and R branch can be represented by a single parabolic equation.

$$v = v_0 + (B_v' + B_v'')m + (B_v' - B_v'')m^2$$

where

$$m = J'' + 1 = 1, 2, 3, \ldots\ldots$$

for lines R(0), R(1), R(2),...

$$m = -J'' = -1, -2, -3, \ldots\ldots$$

for lines P(1), P(2), P(3),...

$$m = 0 \text{ for the band origin.}$$

Franck Condon Principle : A vibrational transition will have a large Franck-Condon factor only if one of the two turning points of a level of one electronic state lies at approximately the same internuclear distance as one of the two turning point of a level of the other electronic state, except in case of $v = 0$ level for which the middle-point of the level rather than the turning point must be substituted.

- The potential curves become increasingly asymmetrical with increasing r. This means that the nuclei spend more time in the extended phase ($r > r_e$) than in the contracted phase ($r < r_e$). Hence according to Franck-Condon principle, transition will be more probable from (or to) the one side than the other.

∴ for a system degraded to the red ($r' > r''$) the right limb of the Condon parabola will be the stronger.

5

Variables Separated

Hydrogen Atom

The hydrogen atom consists of a single electron of mass *m* and charge - *e*, moving about a nucleus of mass *M* and charge + *e* in a potential field *V* (*r*) which corresponds to Coulombian attraction $\frac{e^2}{4\pi\epsilon_0 r^2}$ between the electron and the nucleus. In general, the nucleus is not stationary. Its motion can, however, be taken into account by treating the system as a single particle of (reduced) mass μ moving in the potential field *V* (*r*), where $\mu = mM/(m+M)$. The Schrodinger equation for the motion is

$$\nabla^2 \psi + \frac{8\pi^2\mu}{h^2} [E - V(r)] \psi = 0.$$

Converting the Laplacian operator ∇^2 in spherical coordinates, we get

$$\frac{1}{r^2}\frac{\partial}{\partial r}\left(r^2\frac{\partial\psi}{\partial r}\right) + \frac{1}{r^2 \sin\theta}\frac{\partial}{\partial\theta}\left(\sin\theta\frac{\partial\psi}{\partial\theta}\right) + \frac{1}{r^2\sin^2\theta}\frac{\partial^2\phi}{\partial\phi^2} + \frac{8\pi^2\mu}{h^2}[E - V(r)]\psi = 0. \quad \text{... } (i)$$

This is the Schrodinger equation of hydrogen atom in spherical coordinates. To solve this equation we must separate the radial (r) and angular (θ, ϕ) variables.

Let us substitute

$$\psi\,(r, \theta, \phi) = R\,(r)\; Y\,(\theta, \phi)$$

in eq. (*i*). We get

$$\frac{1}{r^2}\frac{d}{dr}\left(r^2\frac{dR}{dr}\right)Y + \frac{1}{r^2\sin\theta}\frac{\partial}{\partial\theta}\left(\sin\theta\frac{\partial Y}{\partial\theta}\right)R$$

$$+\frac{1}{r^2\sin^2\theta}\frac{\partial^2 Y}{\partial\phi^2}R + \frac{8\pi^2\mu}{h^2}\,[E - V\,(r)]\,RY = 0.$$

Multiplying the entire equation by $\frac{r^2}{RY}$ and rearranging, we get

$$\frac{1}{R}\frac{d}{dr}\left(r^2\frac{dR}{dr}\right) + \frac{8\pi^2\mu r^2}{h^2}\,[E - V\,(r)\,]$$

$$= -\frac{1}{Y}\left[\frac{1}{\sin\theta}\frac{\partial}{\partial\theta}\left(\sin\theta\frac{\partial Y}{\partial\theta}\right) + \frac{1}{\sin^2\theta}\frac{\partial^2 Y}{\partial\phi^2}\right].$$

The left-hand side of this equation is a function of R only, while the right-hand side is a function of Y only. Hence each side must be equal to the same constant. Let this constant be $l\,(l + 1)$. Thus we get a radial equation

$$\frac{1}{R}\frac{d}{dr}\left(r^2\frac{dR}{dr}\right) + \frac{8\pi^2\mu r^2}{h^2}\,[E - V\,(r)] = l\,(l + 1)$$

or

$$\boxed{\frac{1}{r^2}\frac{d}{dr}\left(r^2\frac{dR}{dr}\right) + \left[\frac{8\pi^2\mu}{h^2}\{E - V\,(r)\} - \frac{l(l+1)}{r^2}\right]R = 0,} \quad \dots (ii)$$

and an angular equation

$$-\frac{1}{Y}\left[\frac{1}{\sin\theta}\frac{\partial}{\partial\theta}\left(\sin\theta\frac{\partial Y}{\partial\theta}\right)+\frac{1}{\sin^2\theta}\frac{\partial^2 Y}{\partial\phi^2}\right]=l\,(l+1). \quad \text{... }(iii)$$

The later equation can be further splitted by substituting

$Y(\theta, \phi) = \Theta(\theta)\,\Phi(\phi)$.

Thus we get

$$-\frac{1}{\Theta\,\Phi}\left[\frac{1}{\sin\theta}\frac{d}{d\theta}\left(\sin\theta\frac{d\Theta}{d\theta}\right)\Phi+\frac{1}{\sin^2\theta}\frac{d^2\Phi}{d\phi^2}\Theta\right]=l\,(l+1).$$

Multiplying the entire equation by $\sin^2\theta$ and rearranging, we get

$$\frac{1}{\Theta}\sin\theta\frac{d}{d\theta}\left(\sin\theta\frac{d\Theta}{d\theta}\right)+l(l+1)\sin^2\theta=\frac{1}{\Phi}\frac{d^2\Phi}{d\phi^2}.$$

Again, the two sides of this equation are functions of different variables and so each must be equal to the same constant mt^2 (say). Thus we get

$$\frac{1}{\Theta}\sin\theta\frac{d}{d\theta}\left(\sin\theta\frac{d\Theta}{d\theta}\right)+l(l+1)\sin^2\theta=m_l^2$$

or
$$\boxed{\frac{1}{\sin\theta}\frac{d}{d\theta}\left(\sin\theta\frac{d\Theta}{d\theta}\right)+\left[l(l+1)-\frac{m_l^2}{\sin^2\theta}\right]\Theta=0,} \quad \text{... }(iv)$$

and $-\dfrac{1}{\Phi}\dfrac{d^2\Phi}{d\phi^2}=mi^2$

or
$$\boxed{\frac{d^2\Phi}{d\phi^2}+ml^2\Phi=0.} \quad \text{... }(v)$$

Thus we have broken the Schrodinger equation into three differential equations (*ii*), (*iv*) and (*v*) which are respectively in R alone, in Θ alone and in Φ alone. When we try to find "acceptable' solutions of those equations, the quantum numbers automatically come in.

Hydrogen Atom and Wave Functions

Entrance of Quantum Numbers: The solution of eq. (*v*) is

$$\Phi_{ml}(\phi) = Ae^{iml\,\phi}, \qquad \text{... } (vi)$$

where A is integration constant. It can be shown that $\Phi_{ml}(\phi)$ is a single-valued function of position only when *ml* is a positive or negative integer or zero, that is

$$mi = 0, \pm 1, \pm 2, \pm 3, \dots \dots$$

ml is known as 'magnetic quantum number'.

The solution of eq. (iv) is known to be

$$\Theta_{l,\,ml}(\theta) = N_{l,\,ml}\, p_l^{|ml|}(\cos\theta), \qquad \text{... } (vii)$$

where *Nl, ml* is constant and $Pi^{|mi|}$ is 'associated Legendre polynomial' of degree *l* and order m_l. The solution is found to be acceptable only when *l* is a positive integer, that is,

$$l = 0, 1, 2, \dots\dots\dots$$

l is known as 'orbital quantum number'.

Further the functions $\Theta._{l,mi}(\theta)$ vanish if $|ml|$ exceeds *l*. This means that m_l can take values only upto $\pm\, l$.

For solving the radial equation (*ii*) we must specify *V* (*r*).

In the present case it is $-\dfrac{e^2}{4\pi \in_0 r}$. Then the equation is

$$\frac{1}{r^2}\frac{d}{dr}\left(r^2\frac{dR}{dr}\right) + \left[\frac{8\pi^2\mu}{h^2}\left\{E + \frac{e^2}{4\pi \in_0 r}\right\} - \frac{l(l+1)}{r^2}\right] R = 0$$

or $$\frac{d^2R}{dr^2} + \frac{2}{r}\frac{dR}{dr} + \left[\frac{8\pi^2\mu}{h^2}\left\{E + \frac{e^2}{4\pi \in_0 r}\right\} - \frac{l(l+1)}{r^2}\right] R = 0.$$

The solution of this equation is known to be

$$R_{n,l}(r) = N_{n,l}\left(\frac{2r}{na_0}\right)^i e^{-r/na_0} L_{n+l}^{2l+1}\left(\frac{2r}{na_0}\right), \quad \text{... (viii)}$$

where $N_{n,l}$ is constant and L_{n+l}^{2l+1} is 'associated Legen polynomial' of degree $n = l - 1$ and order $2l + 1$. The parameter a_0 is

$$a_0 = \frac{\epsilon_o h^2}{\pi \mu e^2}$$

is the radius of first Bohr orbit. The solution is found to be acceptable only when $n - l - 1$ is a positive integer. This means that n, I and m_l are restricted to the following values:

$$n = 1, 2, 3, \ldots\ldots\ldots \infty$$

$$l = 0, 1, 2, \ldots\ldots\ldots (n-1)$$

$$ml = 0, \pm 1, \pm 2, \ldots\ldots\ldots \pm 1$$

n is known as 'total quantum number'.

Thus the wave functions of hydrogen atom are given by

$$\psi_{n,l,ml}(r, \theta, \phi) = R_{n'l}(r)\ {}^{\Theta}l_{,ml}(\theta)\ \Phi_{ml}(\phi)$$

where R, Θ and Φ are given by eq. (*viii*), (*vii*), and (*vi*) respectively. The real form of the wave functions of hydrogen atom for the ground state ($n = 1, l = 0.\ m_l = 0$) is given below:

$R_{1,0}(r)$	$\Theta_{0,0}(\theta)$	$\Phi_0(\phi)$	$\psi_{1,0,0}(r, \theta, \phi)$
$\frac{2}{a_0^{3/2}} e^{-r/a_0}$	$\frac{1}{\sqrt{2}}$	$\frac{1}{\sqrt{(2\pi)}}$	$\frac{1}{\sqrt{(\pi a_0^3)}} e^{-r/a_0}$

Wave Mechanical Interpretation of n, I, ml :— The radial wave equation for the hydrogen atom can be solved only when E has a continuous range of positive values, or a discrete set of *negative* values (which correspond to *bound* states), given by

$$E = \frac{\mu e^4}{8 \epsilon_0^2 h^2}\left(\frac{1}{n^2}\right), \quad n = 1,2,3,\ldots$$

These are same as obtained by old quantum theory. Thus n describes the quantisation of electron energy in the hydrogen atom.

Let us now find the interpretation of l and m_l. The wave mechanical operators for the square of the angular momentum and for the z-component of angular momentum are

$$L_2 = -\frac{h^2}{4\pi^2}\left[\frac{1}{\sin\theta}\frac{\partial}{\partial\theta}\left(\sin\theta\frac{\partial}{\partial\theta}\right)+\frac{1}{\sin^2\theta}\frac{\partial^2}{\partial\phi^2}\right] \quad \text{... (ix)}$$

and
$$L_z = -\frac{ih}{2\pi}\frac{\partial}{\partial\phi}. \quad \text{... (x)}$$

The angular eq. (*iii*) can be written as

$$\left[\frac{1}{\sin\theta}\frac{\partial}{\partial\theta}\left(\sin\theta\frac{\partial}{\partial\theta}\right)+\frac{1}{\sin^2\theta}\frac{\partial^2}{\partial\phi^2}\right] Y_{1,mi}(\theta,\phi)$$

$$= -l(l+1)\,Y_{l,mi}(\theta,\phi).$$

Multiplying both sides by $-\frac{h^2}{4\pi^2}$, we get

$$-\frac{h^2}{4\pi^2}\left[\frac{1}{\sin\theta}\frac{\partial}{\partial\theta}\left(\sin\theta\frac{\partial}{\partial\theta}\right)+\frac{1}{\sin^2\theta}\frac{\partial^2}{\partial\phi^2}\right] Y_{l,ml}(1,\phi)$$

Substituting eq. (ix) in it, we have

$$L^2\,Y_{l,ml}(\theta,\phi) = l(l+1)\frac{h^2}{4\pi^2}\,Y_{l,ml}(\theta,\phi).$$

Thus the angular wave functions $Y,_{l,ml}(\theta,\phi)$ of hydrogen atom are also the wave functions of the square of the angular momentum operator with eigenvalues $l(1+1)\frac{h^2}{4\pi^2}$. This means that the electron in the hydrogen atom has an angular momentum given by

$$L = \sqrt{[l(l+1)]}\frac{h}{2\pi},$$

where l = 0, 1, 2,...(n – 1). Thus, like energy, the angular momentum is also quantised, and this quantisation is described by the orbital quantum number l.

Since the magnitude of the electron angular momentum is quantised, its direction is also expected to be quantised with respect to an external magnetic field (space quantisation). Let us consider the Φ-equation (*vi*) :

$$\Phi_{ml}(\phi) = Ae^{imt\phi}$$

Differentiating it, we get

$$\frac{d\Phi_{mt}(\phi)}{d\phi} = Ae^{iml\phi}(i\, mi)$$

$$= i\, mi\, \Phi_{ml}(\phi).$$

Multiplying both sides by $-\frac{ih}{2\pi}$, we get

$$-\frac{ih}{2\pi}\frac{d\Phi_{ml}(\phi)}{d\phi} = \frac{m_l h}{2\pi}\Phi_{ml}(\phi).$$

Substituting eq, (x) in it, we have

$$L_z\,\Phi_{ml}(\phi) = \frac{m_l h}{2\pi}\Phi_{ml}(\phi).$$

Thus the wave function $\Phi_{ml}(\phi)$ of hydrogen is also the wave function of z-component of the angular momentum operator with eigenvalues $\frac{m_l h}{2\pi}$. This means that the component of the electron angular momentum in the field direction (z-axis) is given by

$$L_z = m_l\frac{h}{2\pi},$$

where m_l = 0, ± 1, ± 2,... ..±l. This shows that the number of possible orientations of the angular momentum in a magnetic field is $(2l + 1)$, that is, the angular momentum is space-quantised. Thus the magnetic quantum number m_l describes the space quantisation.

Density of Electron Probability

As far as the exact motion of the electron within the atom is concerned, there is striking difference between old quantum theory and wave mechanics. In the old theory, the electron traces, in general, elliptic orbits whose size, eccentricity and orientation are determined by the quantisation rules. In wave mechanics, on the other hand, the electron cannot be pictured as being in a definite orbit. We can only find the relative probabilities of finding the electron at various locations. This is clearly a consequence *of* the wave nature of the electron.

The electron wave function in a hydrogen atom is given by

$$\psi_{n,l,ml} = R_{n,l}\,\Theta_{l,ml}\,\Phi_{ml}{}'$$

where $R_{n,l,}$ $\Theta_{l,ml}$ and Φ_{ml} describe how $\psi_{n,l,ml}$ varies with r, θ and ϕ respectively. The probability density $|\psi|^2$ may therefore be written

$$|\psi|^2 = |R|^2\,|\Theta|^2\,|\Phi|^2$$

Approximate pictures of $|\psi|^2$ as functions of r, θ and ϕ for some simpler states (for which $l = 0$ and hence $m_l = (0)$ are shown in Fig.

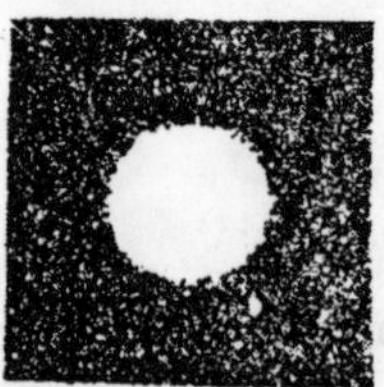 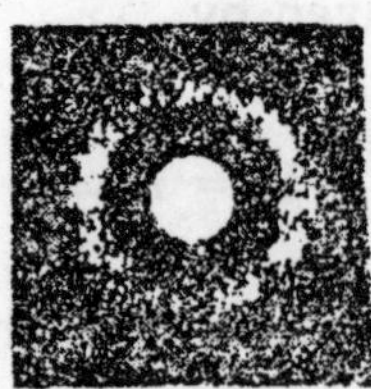 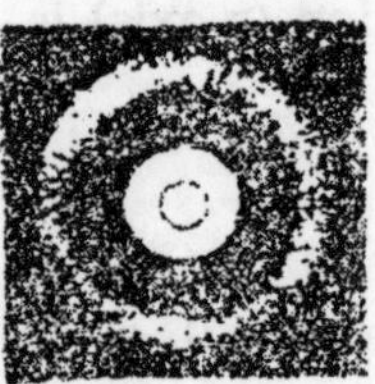

$n = 1$ $n = 2$ $n = 3$

$l = 0$ $l = 0$ $l = 0$

Since we no longer have distinct electron orbits, we speak of *electron clouds* about the nucleus. In the figure, the brightness indicates roughly the density of the electron cloud. (The density of the cloud at a specified point gives the probability of finding the electron at that point).

It is seen that the states with $l = 0$ and $n = 1, 2, 3, \ldots \ldots$ are spherically symmetrical *i.e.* the electron probability density $|\psi|^2$ has the same value at a given r in all directions. (Electron, in another states, however has angular preferences). For $n > 1$, alternate bright and dark rings appear ; the dark rings correspond to nodal surfaces where the electron is never found (zero probability)

Finally, let us see the probability of finding the electron in a hydrogen atom at a distance between r and $r + dr$ from the nucleus. This probability $P(r)$ is given by

$$P(r)\, dr = |R_{n,l}|^2\, 4\pi r^2\, dr,$$

where $4\pi r^2\, dr$ is the volume of the spherical shell whose inner radius is r and outer radius is $r + dr$. In Fig., $P(r)$ has been plotted against r for two states of hydrogen atom corresponding to $n = 1$ and $n = 2$ ($l = 0$ in both cases). It is seen that the principal maxima of the curves closely agree with the radii of the Bohr orbits ($r = a_0, 4a_0, \ldots\ldots$). Thus, whereas the electron is most likely to be found at the locations of Bohr orbits, it has a finite probability of being found elsewhere. Hence, in wave mechanics the electron is, so to speak, smeared out over the whole of space.

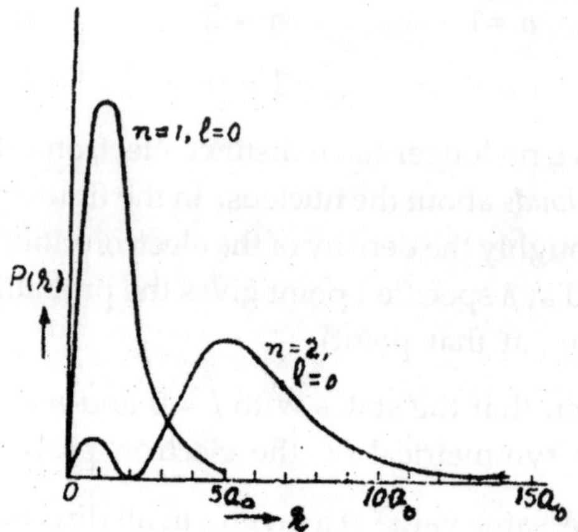

Hydrogen Atom in Normal State

The properties of the hydrogen atom in its normal state (n — I, l = 0, ml = 0) are described by the wave function

$$\psi_{1,0,0} = R_{1,0}\,(r)\,\Theta_{0',0}\,(\theta)\,\Phi_0\,(\phi)$$

$$= \left(\frac{2}{a^{3/2}}\,e^{-r/a_0}\right)\left(\frac{1}{\sqrt{2}}\right)\left(\frac{1}{\sqrt{(2\pi)}}\right)$$

$$= \frac{1}{\sqrt{(\pi a_0^3)}}\,e^{-/a_0}$$

(*i*) The probability distribution function for the electron relative to the nucleus is given by

$$|\psi|^2 = \frac{1}{\pi a_0^3}\,e^{-2r/a_0},$$

which is independent of θ and ϕ. Hence *the normal hydrogen atom is spherically symmetrical.*

(*ii*) The probability that the electron will lie in the volume element $d\tau = r^2\,dr\,\sin\theta\,d\theta\,d\phi$ is given by

$$|\psi|^2\,d\tau = \frac{1}{\pi a_0{}^3}\left(e^{-2r/a_0}\right)\,r^2\,dr\,\sin\theta\,d\theta\,d\phi.$$

Therefore, the probability that the electron will lie between the distances r and $r + dr$ from the nucleus irrespective of its angular distribution is given by

$$P\,(r)\,dr = \frac{1}{\pi a_0{}^3}\left(e^{-2r/a_0}\right) r^2\,dr \int_0^{\pi} \sin\theta\,d\theta \int_0^{2\pi} d\phi$$

$$= \frac{1}{\pi a_0{}^3}\left(e^{-2r/a_0}\right) r^2\,dr\,(2)\,(2\pi)$$

$$= \frac{4}{a_0{}^3}\left(e^{-2r/a_0}\right) r^2\,dr.$$

A plot of radial distribution function $P\,(r)$ against r is shown in Fig. We see that the electron has a large probability of remaining within about 1 Å from the nucleus. This means that the size of the hydrogen atom is about the same as given by Bohr's theory. The distribution function is, however, quite different from that for a Bohr orbit of radius a_0, which would be zero everywhere except at $r = a_0$.

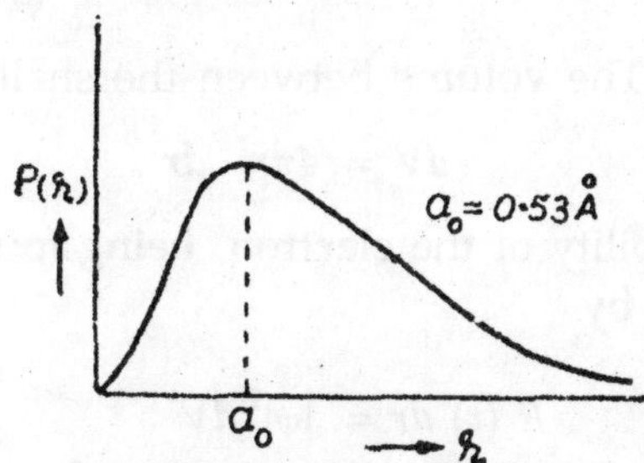

(*iii*) The most probable distance of the electron from the nucleus, *i.e.* the value of r at which $P\,(r)$ is maximum, is obtained by equating $\frac{dP\,(r)}{dr}$ to zero. That is

$$\frac{d}{dr}\left[\frac{4}{a_0^3}\left(e^{-2r/a_0}\right) r^2\right] = 0$$

or $$\frac{4}{a_0^3}\left[e^{-2r/a_0}\,(2r) + r^2\,c^{-2r/a_0}\,(-2/a_0)\right] = 0$$

or $$2r + r^2(-2/a_0) = 0$$

or $$r = a_0 .$$

Thus the most probable distance of the electron from the nucleus in the normal state of hydrogen atom is equal to the Bohr's radius.

Thus, whereas in the old quantum theory the electron is allotted a definite orbit, in wave mechanics it has only a certain probability of being found at a given location. The probability is, however, maximum at points on the Bohr's orbit.

PROBLEMS

1. Consider two hypothetical spherical shells centred on the nucleus of a hydrogen atom with radii r and r + dr. What is the probability P (r) that the electron will lie between these shells, as a function of r ? The motion of the electron in the atom is described by the wave function $\psi \frac{1}{\sqrt{(\pi a_0^3)}} e^{-r/a_0} \cos \omega t$.

Solution: The volume between the shells is

$$dV = 4\pi r^2 \ dr .$$

The probability of the electron being found between the shells is given by

$$P(r)\,dr = |\psi|^2 dV$$

$$= \left(\frac{1}{\pi a_0^3} e^{-2r/a_0} \cos^2 \omega t\right) 4\pi r^2 \ dr.$$

The average probability is obtained by replacing $\cos^2 \omega t$ by its average value which is $\frac{1}{2}$. Thus

$$P(r) = \left\{\frac{1}{\pi a_0^3} e^{-2r/a_0} \left(\frac{1}{2}\right)\right\} 4\pi r^2$$

$$= \frac{2r^2}{a_0^3} e^{-2r/a_0}$$

2. *Show that the degree of degeneracy of the nth energy level is the hydrogen atom in n^2.*

Solution : The energy values given by

$$E = -\frac{\mu e^4}{8 \epsilon_0^2 h^2}\left(\frac{1}{n^2}\right)$$

depend only on n and hence are degenerate with respect to both l and ml (*i.e.* the states having same n but different l's or m_l's have same energy).

For each value of n, l can take n values from 0 to $n - 1$, and for each value of l there are $(2l + 1)$ values of m_l ($= 0, \pm 1, \pm 2, \ldots\ldots \pm l$) Therefore, the degeneracy of the nth energy level is

$$\sum_{i=0}^{n-1} (2l + 1)$$

$$= 1 + 3 + 5 + \ldots\ldots + (2n - 1);$$

a total of n terms

$$= \frac{n}{2} [2(1) + (n - 1)\, 2] = n^2.$$

6

Nuclear Shells

Continuous and Characteristic X-rays : When a beam of X-ray emitted by the target of an X-ray tube is analysed by a Brag's crystal spectrometer, two distinct types of spectra are obtained :

(*i*) a continuous spectrum, and

(*ii*) a line spectrum superposed on the continuous spectrum.

This shows that the X-ray beam from the target is made up of a number of continuous wavelengths of varying intensity; and that the intensity rises sharply at certain definite wavelengths.

Fig. Illustrates roughly the variation of intensity of the beam with wavelength λ when the X-ray tube is operated at a sufficiently high voltage. The continuous spectrum has a definite snort wavelength limit $\lambda_{min,}$ *below* which there is no radiation. This limit is independent of the material of the target and depends on the voltage applied to the X-ray tube. The radiation having the continuous wavelengths forms the `continuous X-rays'.

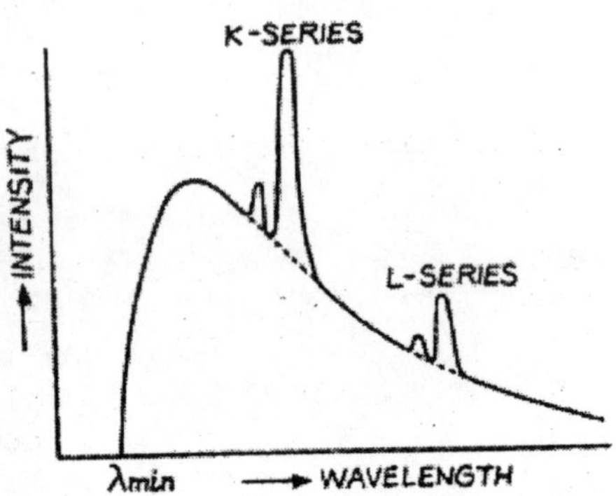

The continuous curve has pronounced intensity peaks at certain wavelengths, indicating the enhanced production of X-rays. The positions of these peaks, however, depend only on the material of the target. They occur in groups known as the *K, L, M, ...,* series. These series of lines are known as the X-ray spectra of the material of the target. The radiation having definite wavelengths which are characteristic of the material of the target forms the 'characteristic X-rays'.

Origin of Continuous X-rays: The emission of continuous spectrum can be explained by the classical theory. When the electrons emitted from the cathode of the X-ray tube strike the target, they are deflected by the strong fields of force of the nuclei of the target atom. Such electrons, according to classical theory, emit pulses of radiation of all possible wavelengths which form the continuous spectrum. The theory, however, fails to explain the occurrence of the short wavelength limit, which requires quantum theory for its explanation.

Short Wavelength Limit and Accelerating Voltage, Duane-Hunt Law: According to the quantum theory, the radiation is emitted (or absorbed) in discrete packets of energy or quanta, called 'photons', the energy of each photon being $h\nu$, where ν is the frequency of the radiation and h is 'Planck's constant'. Now, when an electron of charge e emitted from the cathode is accelerated through the voltage V applied to the X-ray tube, it acquires an amount of kinetic energy eV. When this electron collides an atom of the target, it may lose most or all of its energy, which is re-emitted as an X-ray photon.

For the most favourable collision in which the electron loses the whole of its energy in a single collision with the target atom, an X-ray photon of maximum energy, hv_{max}, is emitted. Thus for an accelerating voltage V, the maximum X-ray photon energy is given by

$$eV = hv_{max}.$$

Consequently, the maximum frequency of X-rays produced by electrons of energy eV is

$$V_{max} = \frac{eV}{h}.$$

A maximum frequency v_{max} corresponds to a minimum wavelength λ_{min}, such that

$$\lambda_{min} = \frac{c}{v_{max}},$$

where c is the speed of light. Therefore, the minimum wavelength of the X-rays for a given voltage V across an X-ray tube is

$$\lambda_{min} = \frac{hc}{eV}.$$

This is Duane-Hunt relation. Putting the valves :

$$h = 6.63 \times 10^{-34} \text{ J-s},$$

$$c = 3.0 \times 10^{3} \text{ m-s}^{-1}$$

and $$e = 1.6 \times 10^{-19} \text{ C, we get}$$

$$\lambda_{min} = \frac{1.24 \times 10^{-6}}{V}\ m = \frac{12400}{V}\ \text{Å}.$$

Thus λ_{min} is inversely proportional to V.

Duane and Hunt experimentally studied the variation of

intensity with wavelength in the continuous spectrum at different accelerating voltages applied to the X-ray tube (Fig.). For each voltage, the intensity curve starts at a particular minimum wavelength, rises rapidly to a maximum and then drops gradually but indefinitely towards the longer wavelengths. The wavelength at which the intensity is maximum depends on accelerating voltage, being shorter for higher voltages. Similarly, the short wavelength limit decreases as the accelerating voltage increases. Infact, Duane and Hunt found that the short wavelength limit varies inversely as the accelerating voltage. Thus, the short wavelength limit of continuous X-rays provides a confirmation of quantum theory.

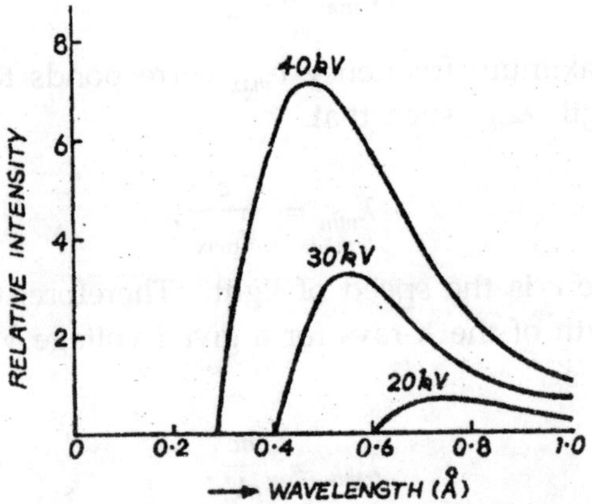

The electron rarely loses whole of its energy in a single collision. More usually it undergoes a sequence of collisions with atoms before coming to rest, thus emitting photons of smaller and smaller energies, *i.e.* of larger and larger wavelengths. This accounts for the production of a continuous range of X-ray wavelengths above a definite limit.

In fact most of the electrons falling upon the target lose their kinetic energy gradually in numerous collisions, thus merely increasing the average kinetic energy of translation of the atoms of the target. The result is that the temperature of the target is increased. Thus most of the kinetic energy of the electron-beam striking the target goes simply into heat. It is only 1% or even less of the electron-energy which is converted into X-rays.

Traits of X-ray Spectrum

When the target of an X-ray tube is struck by energetic electrons, it emits two kinds of X-ray radiation. One of them has a continuous spectrum whose intensity distribution in wavelength depends entirely on the tube voltage. The other consists of distinct spectral lines whose frequencies are characteristic of the material of the target. Hence this line spectrum is called the X-ray spectrum of the material of the target. The radiation forming the line spectrum is called as 'characteristic X-rays'. The frequencies of the characteristic X-rays are, in general, about 1000 times higher than the optical frequencies.

The characteristic *X*-ray spectra are distinguished by their simplicity. Most elements show in their *X*-ray spectra two series of lines known as *K*-series and L-series. Each series contains a small number of lines designated as *K*α, *K*β.......*L*α Lβ.......The wavelengths of *K*-series are generally less than 1 Å, while those of *L*-series are roughly ten times greater. Heavier elements ($Z > 66$) show further series known as M-, *N*-, *O*-, series.

Theory of Kossel

Kossel explained the origin of the characteristic X-ray line spectrum on the basis of the shell structure of the atom. An atom is built up of a central positively-charged nucleus with discrete aggregates of electrons, known as *K*-shell, *L*-shell, *M*-shell, *N*-shell, etc. The maximum number of electrons which the *K, L, M, N,*......shells can hold are 2, 8, 18, 32,....... The electrons in the *K*-shell are attracted by the positive nucleus with the greatest force, and to eject them from the atom maximum energy is required. Lesser energy is required to eject an electron from the L-shell, still lesser from *M*-shell, and so on.

Now, in an X-ray tube an electron emitted from the cathode strikes the target with a tremendous velocity so that it penetrates well inside the atoms of the target. If it ejects an electron from the *K*-shell of the atom, a vacancy is created in the *K*-shell. Immediately an electron from one of the outer shells, say *L*-shell, jumps to the *K*-shell, emitting an *X*-ray photon of energy equal to the energy difference between the two shells. Similarly, if an electron from the *M*-shell jumps to the *K*-shell, X-ray photon of higher energy is emitted. The *X*-ray photons emitted due to the jump of electron from the *L, M, N,...*shells to the *K*-shell give *K*α, *K*β, *K*γ... lines of the *K*-series of the spectrum.

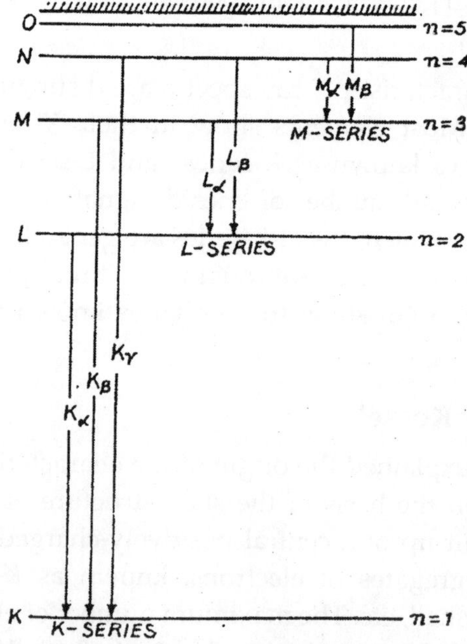

If the electron striking the target ejects an electron from the *L*-shell of the target atom, an electron from the *M, N*...shell jumps to the *L*-shell, so that X-ray photons of lesser energy are emitted. These photons form the *L*-series of the spectrum. Clearly, the frequencies of the L-series are smaller than those of the *K*-series. In a similar way, the formation of *M*-series, *N. series* etc. may be explained. The energy level diagram is shown in Fig. The *K, L,* M,...shells correspond to quantum numbers $n = 1, 2, 3.....$

Whenever a vacancy is created in an inner shell, say *K*-shell, the electron from the next higher shell, the *L*-shell, is most likely to jump to the *K*-shell emitting a line of the *K*-series. The vacancy left by the jumping electron in the *L*-shell is filled by the jump of another electron from a still higher shell, so that a line of *L*-series is emitted. This process continues until the vacancy reaches the outermost shell where it is filled by some free electron. Thus the emission of *K*-series is always accompanied by the emission of *L-, M–, N-* ... series.

Law of Moseley

Moseley, in 1913, made an extensive study of the characteristic X-ray spectra of a number of heavy elements, and observed a simple relationship between them. He found that the spectra of different elements are very similar, and with increasing atomic number Z, the spectral lines merely shift towards shorter wavelengths or higher frequencies (Fig.). He plotted a graph of the K-series between the square-root of frequency $\left(\sqrt{v}\right)$ and atomic number (Z) and found it very approximately to be a straight line (Fig.). He therefore concluded that the square-root of the frequency of a K-line is closely proportional to the atomic number of the element. This is called Moseley's law and may be expressed as

$$\sqrt{v} = k\,(Z - A),$$

where Z is the atomic number of the element, and *k* and *b* are constants for a given transition of the *K*-series. The law is applicable to other series also but with different values of *k* and *b*.

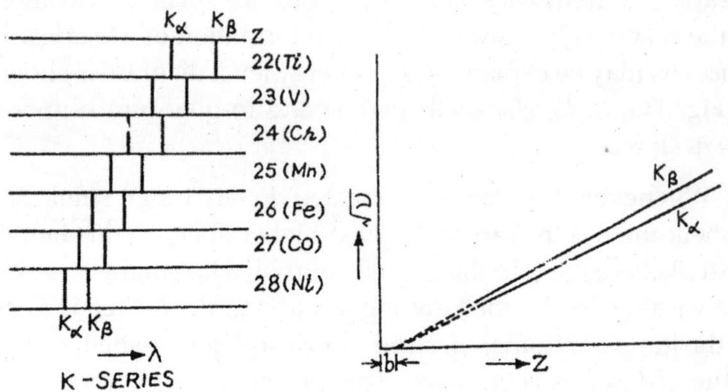

If, however, a graph be plotted between $\sqrt{v}$ and the atomic weight, there is an appreciable departure from straight line. Moseley therefore concluded that it is the atomic number, and not the atomic weight, which is more fundamental with regard to the emission of characteristic X-rays.

Explanation of Moseley's Law from Bohr Theory: At about the same time when Moseley gave his law regarding characteristic X-ray spectra, Bohr presented his theory of the spectrum of the hydrogen atom. It then became evident that Bohr's theory could explain characteristic X-ray spectra, if the atomic number Z was identified as the number of (positive) charges on the nucleus. According to Bohr theory, the wavelength λ of the hydrogen spectral line emitted when an electron jumps from an initial energy level n_i to a final energy level *nf* is given by

$$\frac{1}{\lambda} = RZ^2 \left(\frac{1}{n_{f^2}} - \frac{1}{n_{i^2}} \right), \qquad \text{... } (i)$$

where *R* is Rydberg's constant. Now, in X-ray emission if one of the two *K*-electrons of an atom is removed, an *L*-electron would be attracted by the positive charge of the nucleus in very much the same way as the single electron of a hydrogen atom is attracted by its nucleus. Since the remaining *K*-electron is much nearer the nucleus than the *L*-electron; it produces a screening effect reducing the attraction of the nucleus for the *L*-electron. Thus for the *L*-electron the effective nuclear charge is reduced to (Z – 1) *i.e.* replacing Z by (Z – 1) in the Bohr formula (i), and remembering that when an *L*-electron jumps into a *K*-shell, the quantum number *n* changes from 2 to 1 (*i.e.* $n_t = 2$ and $n_f = 1$) we find that the wavelength of K_α line of an element of atomic number Z is given by

$$\frac{1}{\lambda} = R(Z-1)^2\left(\frac{1}{1^2}-\frac{1}{2^2}\right).$$

The line L_α arises when an *M*-electron jumps into the *L*-shell having a vacancy of one electron (*i.e.* n_i=3 and n_f= 2). Now, the nine electrons remaining in the *K* and *L* shells shield the nucleus during the jump of the electron from the *M* to the *L* shell. However, this shielding is not perfect and the effective Z for the emission of L_α line is reduced only to (Z—7.4). Thus the wavelength of L_α-line is given by

$$\frac{1}{\lambda} = R\,(Z-7.4)^2\left(\frac{1}{2^2}-\frac{1}{3^2}\right).$$

In general, we can write

$$\frac{1}{\lambda} = R\,(Z-b)^2\left(\frac{1}{n_{f^2}}-\frac{1}{n_{i^2}}\right),$$

where *b* is screening constant.

The formula for the frequency *v* (— *c/λ)* can be written as

$$v = Rc\,(Z-b)^2\left(\frac{1}{n_{f^2}}-\frac{1}{n_{i^2}}\right),$$

$$= k^2\,(z-b)^2$$

$$\sqrt{v} = k\,(Z-b),$$

which is Moseley's law.

Importance of Moseley's Law with Regard to the Periodic Table: Mendeleev had arranged the elements in the periodic table in the order of increasing atomic weight. This table had a few anomalies. For example, the properties of Argon, Cobalt and Tellurium were not in agreement with the properties of other elements of their groups.

Moseley's work on X-ray spectra indicated that an element is distinguished by the number of positive charges on the nucleus of its atom *i.e.* by the atomic number Z. Rutherford in the same year showed that the atomic number was the fundamental constant which decided the chemical properties of the atom. Moseley then pointed out that the elements in the periodic table must be arranged in the order of increasing atomic number (instead of atomic weight). From this point of view he changed the positions of certain elements in the Mendeleev's periodic table. For instance, he found that the atomic number of Cobalt ($_{27}Co^{58.9}$) is less than that of nickel ($_{28}Ni^{58.7}$) even though its atomic weight is greater. Therefore he arranged them in the order Co, Ni instead of Ni, Co. Similarly, he placed argon ($_{18}A^{40}$) before potassium ($_{19}K.^{39}$), and tellurium ($_{52}Te^{127.\ 6}$) before iodine ($_{53}\ I^{127}$). When these changes were made, the anomalies of the Mendeleev's table disappeared.

While arranging the elements in the order of increasing atomic number, Moseley had to left certain gaps such as at Z = 43, and 72. He pointed out that the corresponding elements could exist but had not then been discovered. These elements were subsequently discovered : they are technetium and hafnium respectively.

Optical vs X–ray Spectra

The X-ray spectra emitted by different elements are very like one another and much simpler than the corresponding optical (visible) spectra. The main differences between the two spectra are as follows :

(*i*) The frequencies in X-ray spectra are about 1000 times higher than those in optical spectra. This indicates that

X-ray spectra arise from electron transitions among the inner complete shells of very firmly bound electrons. This is also in agreement with the fact that the X-ray spectra of a free element is same as that of its chemical compounds. The optical spectra, on the other hand, arise from transitions of the outermost electrons of the atom.

(*ii*) The X-ray spectra vary smoothly from element to element. This is because the X-ray spectra depend on the binding energies of the electrons in the inner shells. These energies simply increase uniformly from element to element owing to the increasing nuclear charge. The optical spectra, on the other hand, show abrupt periodic changes from element to clement. This is because the electronic structure at the surface of the atom changes periodically from element to element.

(*iii*) The X-ray spectra arise as a result of the complete removal of an electron from an inner shell of the atom. In case of optical spectra the electron from an outer shell is simply raised to a higher level and is not completely removed.

(*iv*) The X-ray emission spectra are entirely different from the X-ray absorption spectra. The former consists of discrete lines while the latter consists of continuous regions bounded by sharp edges. On the other hand, the optical emission spectra are identical to the optical absorption spectra.

It is found that the X-ray spectra of different elements are qualitatively very similar. With increasing atomic number, *i.e.* from element to element the characteristic X-ray lines merely shift towards shorter wavelengths or higher frequencies. There are no abrupt changes from one element to the next. The reason is that X-ray spectra arise from electronic transitions among the inner closed shells of the atom. The electronic arrangement in the innermost shells is regular and essentially the same in all elements (except in very light elements). The nuclear charge, however, increases steadily from element to element and this results in the steady increase in the frequency of the corresponding X-ray lines (Moseley's law).

The optical spectra arise from electronic transitions among the outer shells of the atom. The electronic arrangement in these shells changes periodically and rather abruptly from element to element. Hence the optical spectra also change periodically and abruptly.

Spectrum of X-ray Absorption

Absorption Edges: When a beam of continuous X-rays is passed through the film of a substance and the transmitted beam is sent into an X-ray spectrograph, an X-ray absorption spectrum of that substance is obtained. The nature of this spectrum is quite different from the X-ray emission spectrum of the same substance. None of the emission lines K_α, K_β ...appears in the absorption spectrum. Instead the absorption spectrum shows a continuous region bounded by a sharp edge in the position of the limit of the *K* series. When successive elements in the periodic table are used as absorbers, the edge shifts toward higher frequency (Fig.). Thus these edges are characteristic of the absorber element and are called 'absorption edges'.

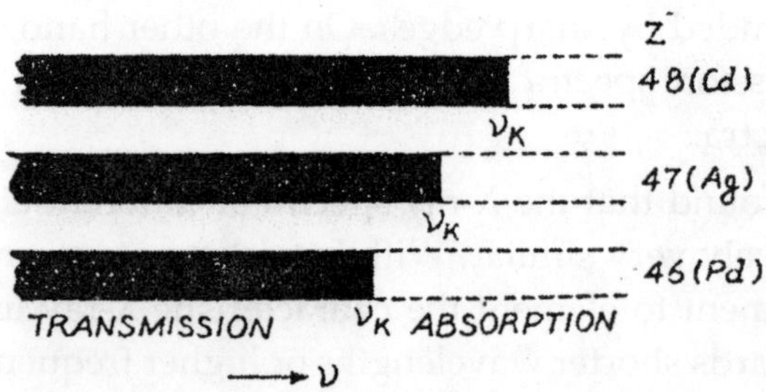

Explanation: These are two features of the X-ray absorption spectrum which are to be explained:

(i) *Absence of Absorption Lines:* The K_α-line (for example) in emission, occurs when an electron drops from the

L-shell to the *K*-shell. In absorption this line would occur only if an electron from the *K*-shell, after absorbing an X-ray photon of the required energy from the continuous beam, could *go* to the *L*-shell. But this does not happen because the *L*-shell has already its full quota of 8 electrons and so it cannot receive any further electron Pauli's exclusion principle). Hence the K_α-line cannot be observed as absorption line. This is true for all the X-ray emission lines that are commonly observed.

(ii) *Appearance of Absorption Edges:* An X-ray photon can, however, be absorbed if it has enough energy to remove an inner electron to infinity where the shells no longer exist. If W_λ- is the energy required to remove a *K*-electron, then a photon of frequency v can eject a *K*-electron provided $v \geq vk$ where

$$hv_K = W_K \cdot$$

Photons of frequency smaller than v_K would be transmitted by the absorbing element.

The black part in Fig. corresponds to the transmission of X-ray photons of all frequencies less then v_K. At the frequency v_K, absorption in the *K*-shell suddenly starts and thereafter this absorption continues for all frequencies greater than v_K. Hence the absorption spectrum is abruptly discontinued at v_K (no blackening of photographic plate beyond v_K).

As Z increases, W_K increases, and so V_k also increases. Hence for successive elements in the periodic table the absorption edge shifts towards higher frequency. The frequency V_k (or the wavelength λ_K) at which absorption in the *K*-shell begins is called the '*K* absorption limit' of the particular element.

Thus *the K-absorpt on edge observed in X-rays, correspond to the photoelectric ejection of an electron from the K-shell.* Absorption beyond the edge indicates that the *K*-electron passes to infinity with the energy E_K given by Einstein's photoelectric equation

($E_K = h\nu — W_K = h\nu - h\nu_K$). Since each element gives only one *K*- absorption edge, we conclude that the *K*-shell has only a single energy level.

Benefits of Multiplicity

When the X-ray absorption spectrum is taken in the *L*-region, again a continuum is observed but now with three absorption edges. This means that the *L*-shell has three energy levels from which electrons can be ejected to infinity. Similarly, M-shell has five energy levels, and so on. Thus X-rays absorption spectrum reveals the multiplicity of atomic shells.

PROBLEMS

1. ***What are the most energetic X-rays emitted in each case when 40-keV electrons bombard (i) Mo target (Z = 42), (ii) Cu target (Z = 29) ? ($e = 1.6 \times 10^{-19}$ C).***

Solution : An X-ray photon of maximum energy is emitted where the electron loses the whole of its energy in a single collision with the target atom. Thus the maximum energy of the emitted X-ray is 40 keV, whichever be the target.

2. ***What is the shortest wavelength of X rays produced in a tube when the applied voltage is 12.4 kV ?***

($h = 663 \times 10^{-34}$ J-s, $c = 3 \times 10^8$ m/s and $e = 1.6 \times 10^{-10}$ C).

Solution : When electrons accelerated through a potential *V* strike a target, the maximum frequency ν_{max} (or minimum wavelength λ_{min}) of the emitted X-ray photon is given by

$$eV = h\nu_{max} = \frac{c}{\lambda_{min}}$$

so that

$$\lambda_{min} = \frac{hc}{cV}.$$

Substituting $V - 12.4$ kV = 12400 volt and the given values of *h*, *c* and *e*, we get

$$\lambda_{min} = \frac{(6.63 \times 10^{-34} \text{ J-s})(3 \times 10^{8} \text{ m/s})}{(1.6 \times 10^{19} \text{ C})(12400 \text{ V})}$$

$$= 1.0 \times 10^{-10} \text{ m}$$

$$= 1.0 \text{ Å}.$$

3. ***An X-ray tube, operating at 66 kV, is producing continuous X-rays. Calculate the shortest wavelength limit of X-rays. (e = 1.6 × 10⁻¹⁰ C. h = 6.6 × l0⁻³⁴ J-s, c = 3.0 × 10⁸ m/s).***

[Ans. 0.19Å]

4. ***What should be the kinetic energy of an electron so that its de Broglie wavelength is same as that of X-rays produced in an X-ray tube operating at 30 kV ? (h = 6.6 × 10⁻³⁴ J.s, c = 3.0 × 10⁸ m/s, e =1.6 ×10⁻⁹ C, m= 9.1 × 10⁻³¹ kg).***

Solution : The (minimum) wavelength of the X-rays produced is

$$\lambda_{min} = \frac{hc}{eV}$$

If K be the kinetic energy and p be the momentum of the electron of mass m, then we have

$$K = \frac{p^2}{2m}$$

But $$p = \frac{h}{\lambda},$$

where λ is the de Broglie wavelength of the electron.

$$\therefore \quad K = \frac{h^2}{2m\lambda^2}$$

Substituting the above value of λ, we get

$$K = \frac{h^2}{2mh^2c^2 / e^2V^2} = \frac{e^2V^2}{2mc^2}$$

$$= \frac{\left(1.6 \times 10^{-19} \text{ coul}\right)^2 \left(30{,}000 \text{ volt}\right)^2}{2\left(9.1 \times 10^{-31} \text{ kg}\right)\left(3.0 \times 10^8 \text{ ms}^{-1}\right)^2}$$

$$= 1.4 \times 10^{-16} \text{ joule.}$$

5. *If the minimum wavelength recorded in the continuous X-ray spectrum of a 100-kV tube is 12.35×10^{-12} meter, calculate the value of Planck's constant. (e = 1.6×10^{-19} coulomb, c = 3×10^8 m/s).*

Solution : The minimum wavelength limit of the X-rays emitted from tube at V volt is given by

$$\lambda_{\min} = \frac{hc}{eV}$$

so that $h = \dfrac{eV}{c} \lambda_{\min}$

$$= \frac{\left(1.6\times 10^{-19} \text{ coul}\right)\left(100 \times 10^3 \text{ volt}\right)\left(12.35 \times 10^{-12} \text{ meter}\right)}{3 \times 10^8 \text{ m/s}}$$

$$= 6.59 \times 10^{-34} \text{ J-s.}$$

6. *An X-ray tube produces X-rays of shortest wavelength of 0.25Å. Calculate the operating voltage. (h = 6.6×10^{-34} J-s, e = 1.6×10^{-19} C, c = 3×10^8 m/s).*

[**Ans.** 49.5 kV]

7. *An X-ray machine produces X-rays of wavelengths 0.1 Å and greater. What voltage does it employ ?*

[**Ans.** 1.24×10^5 volts]

8. *4n X-ray tube with Cu target is operated at 25 kV. The glancing angle for a NaCl crystal for the Cu K_α line is 15.8°. Find the wavelength of this line. Also find the glancing angle for photons at the short wavelength limit.*

(d for NaCl = 2.82 Å. h, c.

Solution: From Bragg's relation,

$$2d \sin\theta = n\lambda$$

For the first order, $n = 1$.

$$\therefore \quad \lambda = 2d \sin\theta$$

$$= 2 \times 2.82 \text{ Å} \times \sin 15.8°$$

$$= 2 \times 2.82 \text{ Å} \times 0.2723 = 1.53 \text{ Å}.$$

The wavelength of X-ray photons at the shortest limit is

$$\lambda_{min} = \frac{hc}{eV} = 0.496 \text{ Å. (Solve)}$$

Let θ' *be* the glancing angle corresponding to $\lambda_{m/n}$. Then

$$\sin\theta' = \frac{0.496 \text{ Å}}{2 \times 1.82 \text{ Å}} = 0.088.$$

$$\therefore \quad \theta' = \sin^{-1}(0.088) = 5.1°.$$

9. If the series limit of the Balmer series for hydrogen is 364.6 nm, calculate the atomic number of the element which gives X-ray wavelengths down to 0.1 nm.

Solution : The Balmer series of hydrogen is given by

$$\frac{1}{\lambda} = R\left(\frac{1}{2^2} - \frac{1}{n^2}\right); \text{ n} = 3, 4, 5 \ldots \ldots \infty.$$

For the series limit ($n = \infty$), we have

$$\frac{1}{\lambda_\infty} = \frac{R}{4}$$

$$\therefore \quad R = \frac{4}{\lambda\infty} = \frac{4}{364.6 \text{ nm}}.$$

The X-ray wavelengths of the *K*-series are given by

$$\frac{1}{\lambda} = R(Z-1)^2\left(\frac{1}{1^2} - \frac{1}{n^2}\right).$$

The minimum wavelength of this series occurs when $n = \infty$ and therefore

$$\frac{1}{\lambda} = R\,(Z-1)^2$$

or $$(Z-1)^z = \frac{1}{\lambda R}$$

Here $\lambda = 0.1$ nm and $R = 4/364.6$ nm. Thus

$$(Z-1)^2 = \frac{364.6 \text{ nm}}{0.1 \text{ nm} \times 4} = 911.5$$

or $$Z - 1 = 30.2.$$

$\therefore$ $$Z \doteq 31, \text{ which corresponds to gallium.}$$

10. How much energy goes into the quantum emitted when an L-electron (i.e. an electron from the n = 2 state) drops into the empty K-state in copper (Z = 29) ?

Given R = 109737 cm^{-1} ; l cm^{-1} = 1.239 × 10^{-4} eV.

Solution : The X-ray wavelength λ emitted, when an $n = 2$ electron drops to the $n = 1$ state, is given by

$$\frac{1}{\lambda} = R\,(Z-b)^2\left(\frac{1}{1^2}-\frac{1}{2^2}\right).$$

For the K–line, $b = 1$. Therefore

$$\frac{1}{\lambda} = \frac{3}{4}R\,(Z-1)^2.$$

Putting the given values of R and Z :

$$\frac{1}{\lambda} = \frac{3}{4}(109737 \text{ cm}^{-1})\,(29-1)^2$$

$$= 6.4525 \times 10^7 \text{ cm}^{-1}.$$

The corresponding energy emitted is

$(6.4525 \times 10^7) \times (1.239 \times 10^{-4}) = 7994.6$ eV.

11. *Estimate the value of the wavelength of K_α line of silver (Z = 47). Given R = 109737 cm^{-1}.*

Solution : From Moseley's law, the wavelength of K_α-line is given by

$$\frac{1}{\lambda_{K\alpha}} = R\,(Z-1)^2.\left(\frac{1}{1^2}-\frac{1}{2^2}\right).$$

$$= \frac{3}{4}\,R\,(Z-1)^2$$

$$= \frac{3}{4}\times 109737\ \text{cm}^{-1}\times(46)^2$$

$$= 1.74\times 10^8\ \text{cm}^{-1}.$$

$$\therefore \qquad \lambda = \frac{1}{1.74\times10^8}\ \text{cm} = \frac{10^8}{1.74\times10^8}\ \text{Å}.$$

$$= 0.57\ \text{Å}.$$

12. *What element has a K_α X-ray line of wavelength 1.785Å? R = 109737 cm^{-1}.*

[**Ans.** Z = 27 (Co)]

13. *If the K_α–radiation of Mo (Z = 42) has a wavelength of 0.71 Å, calculate the wavelength of the corresponding radiation of Cu (Z = 29).*

Solution : From Moseley's law; we have

$$\frac{1}{\lambda_{K\alpha}} \propto (Z-1)^2.$$

$$\therefore \qquad \frac{\lambda_{cu}}{\lambda_{mo}} = \frac{(Z_{mo}-1)^2}{(Z_{cu}-1)^2} = \frac{(41)^2}{(28)^2}$$

But $\lambda_{mo} = 0.71$ Å.

$$\therefore \qquad \lambda c_u = (0.71\ \text{Å})\times\frac{(41)^2}{(28)^2} = 1.52\ \text{Å}.$$

14. *If the K–, L–, M-energy levels of platinum involved in the emission of the K-series of X-rays of this element are*

78000, 12000 and 3000 eV respectively, calculate the wavelengths of the K_α and K_β lines of platinum. ($h = 6.63 \times 10^{-34}$ joule-sec, $c = 3.0 \times 10^8$ meter/sec, 1 eV = 1.6×10^{-19} joule).

Solution : The K_α–line is emitted when an electron drops from the *L*-shell to the *K*-shell. Hence the energy of a K_α-photon is the difference in energies of these two shells, *i.e.*

$$h\nu_{K\alpha} = 78000 - 12000 = 66000 \text{ eV}$$

$$= 66000 \times (1.6 \times 10^{-19}) \text{ joule.}$$

$$\therefore \quad v_{k\alpha} = \frac{6600 \times \left(1.6 \times 10^{19}\right)}{6.63 \times 10^{-34}} \text{ sec}^{-1}.$$

The corresponding wavelength is

$$\lambda_{K\alpha} = \frac{c}{v_{K\alpha}}$$

$$= \frac{\left(3.0 \times 10^{8}\right) \times \left(6.63 \times 10^{-34}\right)}{66000 \times \left(1.6 \times 10^{-19}\right)}$$

$$= 1.88 \times 10^{-11} \text{ meter}$$

$$= 0.188 \text{ Å}.$$

The K_β-line is emitted when an electron drops from the *M*-shell to the *K*-shell. Thus

$$hv_{K\beta} = 78000 - 3000 = 75000 \text{ eV}.$$

$$\therefore \quad \lambda_{K\beta} = \frac{\left(3.0 \times 10^{8}\right) \times \left(6.63 \times 10^{-34}\right)}{75000 \times \left(1.6 \times 10^{-19}\right)}$$

$$= 1.65 \times 10^{-11} \text{ meter} = 0.165\text{Å}.$$

15. If the K-absorption limit of Uranium is 0.107 Å, find the minimum potential difference required across an X-ray tube to excite the K-series ?

Solution : If v_K be the frequency of the K-absorption limit for the target material, the critical voltage V_K required is given by

$$eV_K = hv = \frac{hc}{\lambda_K}$$

$$\therefore \qquad Vk = \frac{hc}{e\lambda_K}$$

Now $h = 6.63 \times 10^{-34}$ joule-sec,

$c = 3.0 \times 10^{8}$ meter/sec,

$e = 1.6 \times 10^{-19}$ coulomb

and $\lambda_K = 0.107$ Å $= 0.107 \times 10^{-10}$ meter.

$$\therefore \qquad V_K = \frac{(6.63 \times 10^{-34}) \times (3.0 \times 10^{8})}{(1.6 \times 10^{-19}) \times (0.107 \times 10^{-10})}$$

$$= 116 \times 10^{3} \text{ volts} = 116 \text{ kV}.$$

7

Series Relationship

The Spectrum

Intercorrection of Series

Spectrum: When a beam of white light is passed through a prism (or grating) it breaks up into the beams of constituent colours. The different coloured beams, when focussed on a screen by a converging lens, form an array of colours on the screen, This is called the 'spectrum' of light.

There are two main types of spectra : (*i*) 'emission spectra' and (*ii*) 'absorption spectra'.

When light coming directly from a source is allowed to enter a prism or grating spectroscope, an 'emission spectrum' of the source is observed in the spectroscope. On the other hand, when light from a source showing a continuous emission spectrum is passed through an absorbing material and then into the spectroscope, an' absorption spectrum of the material is observed.

Emission and absorption spectra are further classified according to their appearance. There are three classes :

Continuous Spectra : When the emitting source is an incandescent solid, or liquid, such as a lamp filament or a gas at high pressure, the spectrum is continuous containing all colours (wavelengths) from red to violet. Its appearance is like an unbroken luminous band of light in which the colour changes gradually from point to point but without any sharp boundary. In this spectrum the intensity is maximum at a certain point and decreases on both sides of it. The point of maximum intensity shifts towards the violet end of the spectrum as the temperature of the source increases.

Line Spectra : If the light source is a low-pressure gas (as in a discharge tube), flame, arc or spark, the spectrum is discontinuous, showing a number of sharp bright-coloured lines. These lines are the images of the slit of the spectroscope formed by lights of different colours. The entire series of images is called a 'line spectrum.' The different lines differ in intensity and nature. Some are sharp, some are sharp on one side and diffused on the other, and others are diffused on both sides.

The line spectrum is the characteristic of the atom or the ion. It means that a particular atom or ion always gives the characteristic set of spectral lines, and no two atoms or ions can give the Atomic and Nuclear same spectral line. For example, sodium atom gives two intense yellow lines called D_1 and D_2 lines.

Band Spectra : Such spectra are obtained by the radiation from gas molecules such as oxygen (O_2), nitrogen (N_2), cyanogen (CN), etc. They consist of illuminated regions, called 'bands', separated by dark spaces. With a high-resolving instrument each band is seen to consist of very fine lines which become closer and closer on one side of the band until they coincide. This side thus has a sharp and bright edge called the 'head' of the band.

Absorption Spectra : The absorption spectra are obtained when the absorbing substance is placed between a source emitting a continuous spectrum and the slit of the spectroscope.

In such cases certain colours (wavelengths) are absorbed by the substance. Hence the spectrum is found to consist of dark lines or bands against a bright background. An example of such spectra is sun's spectrum. It is a line absorption spectrum. It was studied in detail by Fraunhoffer who named the absorption lines as *A,B,C,D......* These lines inform us about the elements which are present around the sun.

Similarly, when an intense beam of continuous white light is passed through sodium vapour and then sent into a spectroscope, we obtain two dark lines on a continuous background in the same positions as the yellow D_1 and D_2 lines in the sodium emission spectrum. If the sodium vapour is replaced by Iodine vapour (l_2), an absorption band spectrum of I_2 molecule is obtained.

Series Relationship in Atomic Spectra—The atomic spectra consist of a large number of lines. A quantitative experimental study of these spectral lines was made in the second half of the 19th century when several regularities were observed in the pacings of the lines. For example, in 1870, Liveing and Dewar noticed that the spectral lines of various elements could be grouped into distinct 'series'. *In each series the spacing and intensity of lines decrease regularly towards shorter wavelengths,* until it becomes impossible to distinguish the individual lines. The point at which the lines of the series finally converge is called the 'series limit'. The various series in complicated spectra overlap.

The simplest atomic spectrum is that of hydrogen. Its visible part consists of a single series which was first observed by Balmer in 1885. This is called the 'Balmer series' of hydrogen. Its first line having longest wavelength of 6563A is named H a, the next Hβ , and so on, the series limit reaching at 3646A. Besides this, there is a series of lines in the ultraviolet part of the hydrogen spectrum which is known as 'Lyman series; and three series in the infrared part which are known as 'Paschen series', 'Brackett series' and 'Pfund series'.

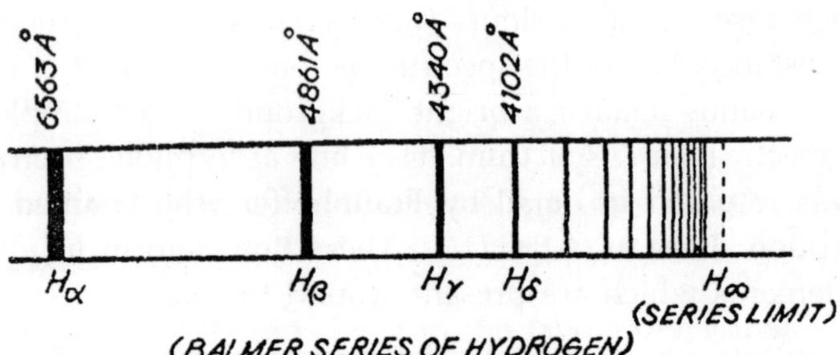

(BALMER SERIES OF HYDROGEN)

Balmer discovered a formula for the wavelengths of all the lines of the Balmer series. His formula is

$$\lambda = 3646 \frac{n^2}{n^2 - 4} \text{ Å} ; n = 3, 4, 5, \ldots \ldots$$

$n = 3$ gives the wavelength of H_α line, $n = 4$ gives H_β line, ... and $n = \infty$ gives the series limit.

Subsequently, Rydberg found that Balmer's formula was a special case of a more general formula which is as follows:

$$\bar{\nu} = \frac{1}{\lambda} = R_{\mathrm{H}}\left(\frac{1}{m^2} - \frac{1}{n^2}\right),$$

where $\bar{\nu}$ is wave number (reciprocal of wavelength), m and n are positive integers ($n > m$) and R_H is the Rydberg constant for hydrogen. Its value is 1.097×10^7 meter[1]. To obtain formula for Lyman series we set $m = 1$ and $n = 2, 3, 4, \ldots \ldots$; for Balmer series $m = 2$ and n —3, 4, 5,......; and so on. Thus

$$\bar{v} = R_H \left(\frac{1}{1^2} - \frac{1}{n^2}\right) ; n = 2, 3, 4, \ldots \ldots \quad \text{(Lyman)}$$

$$\bar{v} = R_H \left(\frac{1}{2^2} - \frac{1}{n^2}\right) ; n = 3, 4, 5, \ldots \ldots \quad \text{(Balmer)}$$

$$\bar{v} = R_H \left(\frac{1}{3^2} - \frac{1}{n^2}\right) ; n = 4, 5, 6, \ldots \ldots \quad \text{(Paschen)}$$

$$\bar{v} = R_H \left(\frac{1}{4^2} - \frac{1}{n^2}\right) ; n = 5, 6, 7, \ldots \ldots \quad \text{(Brackett)}$$

$$\overline{v} = R_H \left(\frac{1}{5^2} - \frac{1}{n^2} \right) ; n = 6, 7, 8, \ldots\ldots \qquad \text{(Pfund)}$$

We see that the wave numbers of hydrogen lines can be expressed as differences of two terms of the form $\frac{R_H}{n^2}$.

The next simplest spectra are of the 'monovalent' atoms of alkali metals Li, Na, K, etc. The lines in the spectrum of an alkali atom can be grouped into four distinct series ; a 'principal series' of intense lines, a 'sharp series' of fine lines, a 'diffuse series' of comparatively broader lines and a 'fundamental series' which lies in the infrared region. The sharp and diffuse series lie in the visible part and converge to a common limit. Rydberg represented the lines of a particular series by the formula

$$\overline{v} = Z^2 R_A \left[\frac{1}{(m - \Delta_1)^2} - \frac{1}{(n - \Delta_2)^2} \right].$$

where R_A is the Rydberg constant for a particular element *A*, *Z* is the atomic number, *m* and *n* are positive integers, and Δ_1 and Δ_2 are constants for the particular series. Actually each line of an alkali spectrum is a close doublet. Again we find that the wave numbers of the lines can be expressed as differences of two terms like $\frac{Z^2 R_A}{(m - \Delta)^2}$

After alkali spectra, next in complexity are the spectra of 'divalent' atoms of alkalineearths Be, Mg, Ca. In a typical alkaline-earth spectrum we can distinguish two distinct systems of lines : a system of singlets and a system of triplets. Each system has a principal series, a sharp series, a diffuse series and a fundamental series.

As we proceed to atoms having several valence electrons, the spectra become more complex and the groupings of lines into series become less pronounced. Still regularities can be observed in complex spectra, and it is possible to express the wave number of any spectral line as the difference of two terms.

Law of Rydberg-Ritz Combination

This principle states that *"the wave numbers of spectral lines can be expressed by the difference of spectroscopic terms in such a way that other differences of those terms give also the wave numbers of lines in the same spectrum".*

Suppose in a spectrum the wave numbers of two lines are given as

$$\overline{v}_a = T_2 - T_3 \text{ and } \overline{v}_d = T_1 - T_4,$$

then lines of the following wave numbers are also expected in the same spectrum;

$$\overline{v}_b = T_2 - T_4 \text{ and } \overline{v}_0 = T_1 - T_3 .$$

This means that constant differences exist between the wave numbers:

$$\overline{v}_b - \overline{v}_a = \overline{v}_d - \overline{v}_c$$

$$\overline{v}_c - \overline{v}_a = \overline{v}_d - \overline{v}_b$$

Displacement Law: According to this law 'the spectrum of any neutral atom of atomic number Z closely resembles the spectrum of the singly ionised atom of atomic number Z + 1. For example, the spectrum of He + (Z = 2) closely resembles the spectrum of H (Z = l) and can be represented by a similar formula,

$$\overline{v} = 4 R_{He} \left[\frac{1}{m^2} - \frac{1}{n^2}\right].$$

Other elements deprived of all but one electron, also produce hydrogen-like spectra which can be represented by the general formula

$$\bar{v} = Z^2 R_A \left[\frac{1}{m^2} - \frac{1}{n^2}\right].$$

In a similar way, the spectrum of a singly-ionised alkaline-earth atom resembles the spectrum of an alkali atom. From this we conclude that it is the number of valence electrons in an atom which determines the qualitative character of the spectrum of that atom.

PROBLEMS

1. The wavelength of a yellow line of sodium is 5896 Å. Its wave number and frequency are calculated as follows.

Solution : The wave number $\bar{v}$ is the reciprocal of wavelength λ, that is,

$$\bar{v} = \frac{1}{\lambda}.$$

Here $\lambda = 5896 \text{ Å} = 5896 \times 10^{-8} \text{ cm}.$

$$\therefore \quad \bar{v} = \frac{1}{5896 \times 10^{-8} \text{ cm}}$$

$$= 16960 \text{ per cm.}$$

The frequency v is related to the wavelength A by

$$c = v\lambda,$$

where c is the speed of light. Thus

$$v = \frac{c}{\lambda} = \frac{3 \times 10^8 \text{ cm/sec}}{5896 \times 10^{-8} \text{ cm}}$$

$$= 5.1 \times 10^{12} \text{ sec}^{-1}$$

8

Spectrum Emission

The spectrum of hydrogen atom consists of a number of lines. These lines have been grouped into a number of 'series'. The lines in each series are such that *their separation and intensity decrease regularly towards shorter wavelengths,* converging to a limit called the 'series limit'.

The wavelengths in each series can be given by a simple empirical formula. The first such spectral series was observed by Balmer in 1885 and is called the Balmer series of hydrogen. The first line with the longest wavelength (6563 A) is named $H\alpha$, the next $H\beta$, and so on. The series limit lies at 3646 A, beyond which is a faint continuous spectrum. Balmer's formula for the wavelengths of the series is

$$\frac{1}{\lambda} = R\left(\frac{1}{2^2} - \frac{1}{n^2}\right), \quad n = 3, 4, 5, \ldots \ldots \quad \text{(Balmer)}$$

The quantity R is called the 'Rydberg constant' and has the value

$$R = 1.097 \times 10^7 \text{ meter}^{-1}.$$

The H_α line corresponds to $n = 3$, the H_β line to $n = 4$, and so on. The series limit corresponds to $n = \infty$, so that it occurs at a wavelength of $4/R$, in agreement with experiment.

The Balmer series contains only those spectral lines which fall in the visible part of the hydrogen spectrum. The lines falling in the ultraviolet and infrared parts form other series. The lines in the ultraviolet form the Lyman series whose wavelengths are given by

$$\frac{1}{\lambda} = R\left(\frac{1}{1^2} - \frac{1}{n^2}\right), \; n = 2, 3, 5, \dots \dots \qquad \text{(Lyman)}$$

In the infrared, three spectral series have been observed whose lines have the wavelengths given by the formulas

$$\frac{1}{\lambda} = R\left(\frac{1}{3^2} - \frac{1}{n^2}\right), \; n = 4, 5, 6, \dots \dots \qquad \text{(Paschen)}$$

$$\frac{1}{\lambda} = R\left(\frac{1}{4^2} - \frac{1}{n^2}\right), \; n = 5, 6, 7, \dots \dots \qquad \text{(Brackett)}$$

$$\frac{1}{\lambda} = R\left(\frac{1}{5^2} - \frac{1}{n^2}\right), \; n = 6, 7, 8, \dots \dots \qquad \text{(Pfund)}$$

Bohr's Theory of Hydrogen Spectrum : The existence of sharp spectral lines cannot be explained from the classical electromagnetic theory. Bohr explained it by applying Planck's quantum hypothesis to the Rutherford's atomic model. The Rutherford's atom consists of a central massive nucleus containing the positive charge of the atom, and the electrons move round the nucleus in circular planetary orbits.

The centripetal force required for the orbital motion is provided by electrostatic attraction between the positively-charged nucleus and the negatively-charged electron.

Bohr Proposed Two Postulates

(i) An electron can move *only* in those orbits for which the angular momentum *L* of the electron is an integral multiple of $h\backslash 2\pi$ where *h* is Planck's constant. (Thus Bohr quantised the angular momentum of the electron.) The electron moving in any of the permitted orbits does not radiate energy in spite of its acceleration towards the centre of the orbit. The atom, therefore, is said to exist *in* a *stationary* state.

(ii) The emission (or absorption) of radiation by the atom lakes place when an electron jumps from one permitted orbit to another. The radiation is emitted (or absorbed) as a single quantum (photon) whose energy hv is equal to the difference in energies of the electron in the two orbits involved. Thus if *E*i be the energy of the initial orbit of the electron and *Ef* that of the final orbit, then we have

$$hv = E_i - E_f$$

where v is the frequency of the emitted (or absorbed) radiation.

Let e, m and v be the charge, mass and velocity of the electron (measured in coulomb, kg and meter/sec respectively) and r the radius of the orbit measured in meter. The positive charge on the nucleus is Ze, where Z is the atomic number (Fig.). In case of hydrogen Z=l, so that positive charge on nucleus is e. As the centripetal force is provided by the electrostatic attraction, we have

$$\frac{mv^2}{r} = \frac{1}{4\pi\epsilon_0}\frac{e^2}{r^2}$$

or $$mv^2 = \frac{e^2}{4\pi\epsilon_0 r}. \quad \text{... } (i)$$

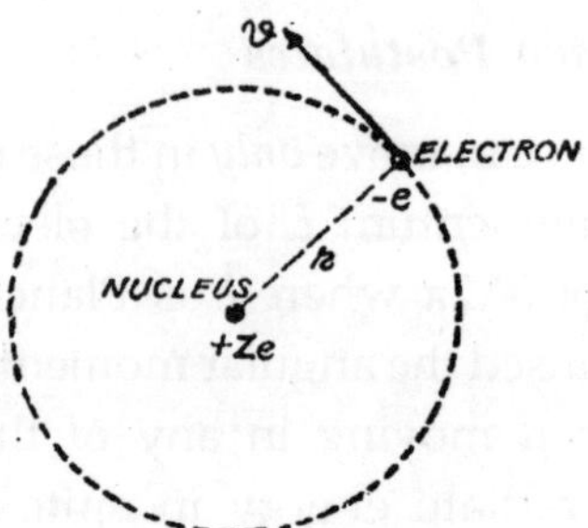

From the first postulate, the angular momentum of the electron is given by

$$L = mvr = n\frac{h}{2\pi},$$

where n called as 'quantum number' = 1, 2, 3,...

Squaring eq. (*ii*) and dividing by eq. (*i*), we get

$$\boxed{r = n^2\frac{h^2\epsilon_o}{\pi me^2},} \quad n = 1, 2, 3, \ldots \qquad \ldots (iii)$$

This is the expression for the radius of the permitted orbits.

Now, the energy E of the electron in an orbit is the sum of kinetic and potential energies. The kinetic energy of the electron

$$K = \frac{1}{2}mv^2 = \frac{e^2}{8\pi\epsilon_0 r}. \qquad \text{[form eq. (}i\text{)]}$$

The potential energy at a distance r from the nucleuses equal to the work done in removing the electron from r to infinity *against the electrostatic attraction* $\left(-\frac{e^2}{4\pi\epsilon_0 r^2}\right)$, and is given by

$$U = \int_r^\infty -\frac{e^2}{4\pi\epsilon_0 r^2}\,dr = \frac{1}{4\pi\epsilon_0}\left[\frac{e^2}{r}\right]^\infty = -\frac{e^2}{4\pi\epsilon_0 r}.$$

Hence the total energy of the electron is

$$E = K + U = \frac{e^2}{8\pi\epsilon_0 r} - \frac{e^2}{4\pi\epsilon_0 r} = -\frac{e^2}{8\pi\epsilon_0 r}.$$

Substituting for r from eq. (iii), we get

$$\boxed{E = -\frac{me^4}{8\epsilon_0^2 h^2}\left(\frac{1}{n^2}\right);} \quad n = 1, 2, 3, \ldots \qquad \ldots(iv)$$

This is the expression for the energy of the electron in the *n*th orbit. We see that it is negative.

Let E_i and E_f be the energies of the election corresponding to the initial (higher) and final (lower) orbits of the excited atom. Then we have

$$E_i = -\frac{me^2}{8\epsilon_0^2 h^2}\left(\frac{1}{n_i^2}\right);$$

and $$Ef = -\frac{me^4}{8\epsilon_0^2 h^2}\left(\frac{1}{n_f^2}\right)$$

where n_i and n_f are the corresponding quantum numbers. The energy difference between these states is

$$E_i - E_f = \frac{me^4}{8\epsilon_0^2 h^2}\left(\frac{1}{n_f^2} - \frac{1}{n_i^2}\right).$$

Hence, from Bohr's second postulate, the frequency v of the emitted photon is

$$v = \frac{E_i - E_f}{h}$$

$$= \frac{me^4}{8\epsilon_0^2 h^3}\left(\frac{1}{n_f^2} - \frac{1}{n_i^2}\right).$$

The corresponding wavelength A is given by

$$\frac{1}{\lambda} = \frac{v}{c} = \frac{me^4}{8_0^{\epsilon^2} ch^3}\left(\frac{1}{n_f^2} - \frac{1}{n_i^2}\right)$$

This equation indicates that, *since n_i and nf can take only integral values, the radiation emitted by excited hydrogen atoms should contain certain discrete wavelengths only.*

The value of the constant *term* $\frac{me^4}{8_0^{\epsilon^2} ch^3}$ comes out to be the same as the Rydberg constant R in the Balmer's empirical formula. Thus we have

$$\frac{1}{\lambda} = R\left(\frac{1}{n_f^2} - \frac{1}{n_i^2}\right).$$

Hydrogen Spectrum

When the hydrogen atom gets sufficient energy from outside by some means, the electron from an inner orbit of lower energy goes up to an outer orbit of higher energy. This excited state of the atom lasts for about 10 ^{8} second after which the electron jumps back to an inner orbit. In jumping down, it emits the difference in energy between two orbits as electromagnetic radiation. If the electron jumps from an orbit n_i to an orbit n_f, the wavelength of the emitted radiation will be

$$\frac{1}{\lambda} = R\left(\frac{1}{n_f^2} - \frac{1}{n_i^2}\right).$$

It is found that for

$n_f = 1$, $n_i = 2, 3, 4, \ldots$ we obtain Lyman series,

$n_f = 2, n_i = 3, 4, 5, \ldots$ we obtain Balmer series,

$n_f = 3, n_i = 4, 5, 6, \ldots$, we obtain Paschen series,

$n_f = 4,\ n_i = 5, 6, 7, \ldots$ we obtain Brackett series,

$n_f = 5, n_i = 6, 7, 8, \ldots$ we obtain Pfund series.

The corresponding energy level diagram is shown in Fig. below. The top horizontal line represents zero energy *i.e.* energy of the electron outside the atom ($n = \infty$). The other horizontal lines represent energies of different orbits given by the formula

$$E = -\frac{me^4}{8\epsilon_0^2 h^2}\left(\frac{1}{n^2}\right).$$

The arrows ending at the lines $n = 1, 2, 3, 4$ and 5 represent the transitions responsible for the Lyman, Balmer, Paschen, Brackett and Pfund series respectively.

Shortcomings of Bohr's Theory : Bohr's theory although very successful in explaining the hydrogen spectrum and giving valuable information about atomic structure, has the following shortcomings :

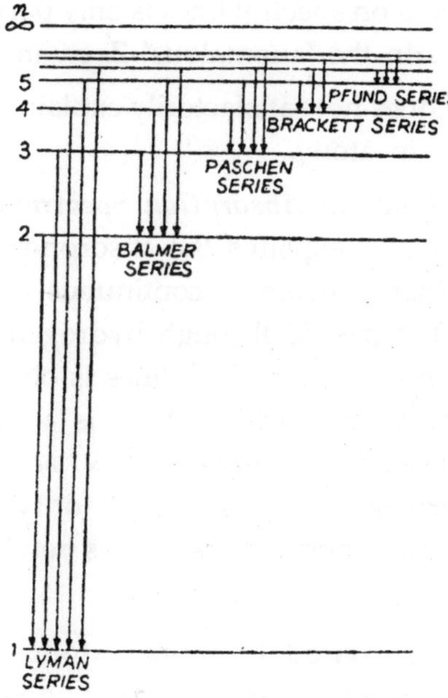

1. An individual line of hydrogen spectrum, when examined under a high-resolving spectroscope is found to be accompanied by a number of faint lines. This 'fine structure' of spectral lines cannot be explained by Bohr's theory as such. It can, however, be explained when
   (i) the relativistic variation in the mass of the electron and
   (ii) the electron 'spin' are taken into account.
2. Bohr's theory cannot explain the variation in intensity of the spectral lines of an element. The intensity can be explained by quantum mechanics.
3. The theory is only applicable to one-electron atoms such as hydrogen, hydrogen isotopes, singly-ionized helium, doubly-ionised lithium, etc. It does not explain the spectra of complex atoms.
4. The success of Bohr's theory in explaining the effect of magnetic field on spectral lines is only partial. The theory cannot explain the 'anomalous' Zeeman effect'.
5. The theory does not satisfactorily explain the distribution of electrons in atoms.

No Balmer Lines in Absorption Spectrum of Hydrogen: The Bohr theory also explains the absorption line spectrum of hydrogen. When a beam of continuous light (containing all wavelengths) is passed through hydrogen and then sent into a spectrograph, a set of dark lines is obtained. In terms of quantum theory, the incident light is a beam of quanta (photons) of all sorts of energies. Now, according to Bohr theory, the hydrogen atoms absorb only those quanta whose energies correspond to transitions between its discrete energy levels.

The resulting excited hydrogen atoms re-radiate the absorbed energy almost at once but these photons come off in random directions with only a few in the same direction as the original beam of continuous light. The dark lines in the absorption spectrum are therefore never completely black. Obviously the absorption lines will have exactly the same frequencies as the emission lines.

Now it is found that all the emission lines of hydrogen spectrum do not appear in the absorption spectrum. The reason is that normally the atom is always in the ground state $n = 1$. Therefore, absorption transitions can only occur from $n = 1$ to $n>1$. Hence lines of only the Lyman series can appear in absorption spectrum. To obtain Balmer series in absorption, the atom must initially be in the state $n = 2$, because Balmer lines require transitions from $n = 2$ to $n = 2$. Since atoms are usually in the ground state, Balmer lines are not obtained in absorption.

Assumptions of Planck, Einstein and Bohr: Planck had assumed that the atoms of a hot body behave as oscillators and have discrete (quantised) energies. They do not emit radiant energy continuously, but only in 'jumps' or 'quanta'. Planck, however, still maintained that radiation propagates continuously through space as electromagnetic waves.

Einstein, in order to explain the photoelectric effect, went a step ahead. He proposed that the radiation not only is emitted as quantum at a time, but also propagates as individual quanta (photons). He thus treated the propagation of radiation as particle propagation rather than wave propagation.

Bohr, in order to explain the hydrogen spectrum, adopted the Planck's quantum hypothesis that the radiation is emitted discontinuously from the atom. He started with the quantisation of the angular momentum of the electron in the orbit which ultimately results in the quantisation of energy of the atom.

Hydrogen Orbits and Energy

An electron revolving in a hydrogen orbit has a negative potential energy by virtue of its attraction towards the nucleus, and also kinetic energy (which is positive) by virtue of its motion. The potential energy is greater in magnitude than the kinetic energy, so that the net energy is negative. The 'negative' energy signifies that the electron cannot escape from the atom. A positive energy for a nucleus-electron combination would mean that electron is not bound to the nucleus. Such a combination cannot constitute an atom.

Absorption of a Photon of Energy Greater than the Binding Energy : The binding energy of the hydrogen atom is the energy which must be supplied to it in order to remove its electron from the lowest state (level) to a state of zero total energy. It is numerically equal to the energy of the lowest state (: 136 eV).

The highest energy level of the atom corresponds to n=∞ so that £=0. Above this are the energy states of the system consisting of an unbound electron plus the ionized atom. The total energy of an unbound electron is not quantised and is positive. Thus any energy $E > 0$ is possible for the electron, and the energy states form a continuum. When the hydrogen atom receives a photon of *any* energy which is greater than its binding energy of 13.6 eV, it absorbs the photon and the electron of the atom passes from its discrete ground state to a continuum state.

Bohr Theory as Corrected for Nuclear Mass: In Bohr theory we assume that the nucleus of the hydrogen atom is infinitely heavy compared with electron, and so it remains stationary while the electron revolves around it. In fact, the nucleus has a *finite* mass, and both the electron and the nucleus revolve about their common centre of mass *C* (Fig.) with a common angular velocity ω. Let us make correction in Bohr theory for the finite mass (motion) of the nucleus.

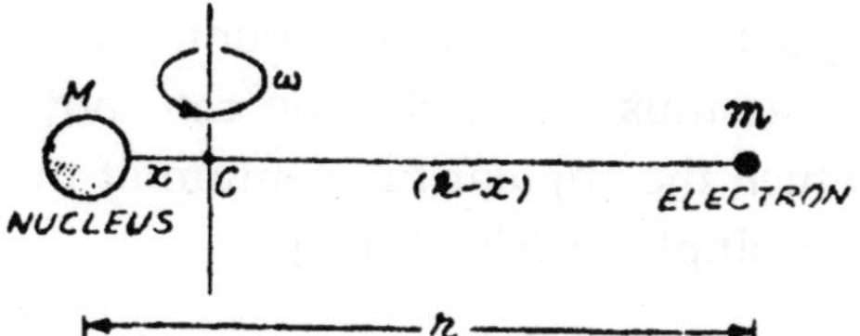

Let m be the mass of electron, M_H that of hydrogen nucleus, and r the distance between them. Let x be the distance of the nucleus from the common centre of mass C. The distance of the electron will be (r—x). Since the system is in equilibrium, the moments of M_H and m about C will be equal. That is

$$M_H\, x = m\,(r - x).$$

Solving we get: $x = \dfrac{mr}{M_H + m}$

and $$r - x = \frac{M_H\, r}{M_H + m}$$

The total angular momentum of the atom about the centre of mass is

$$L = M_H\, x^2\, \omega + m\,(r - x)^2\, \omega$$

$$= M_H \frac{m^2 r^2 \omega}{(M_H + m)^2} + m \frac{M_H^2\, r^2 \omega}{(M_H + m)^2}$$

$$= M_H \frac{m r^2 \omega}{(M_H + m)^2} (m + M_H)$$

$$= \left(\frac{M_H\, m}{M_H + m}\right) r^2 \omega$$

$$= \mu\, r^2\, \omega,$$

where $\mu \left(\dfrac{M_H\, m}{M_H + m}\right)$ is called the 'reduced mass' of the electron, because its value is slightly less than m.

Taking nuclear motion into account, Bohr's first postulate will be read as thus : "An electron can move only in those orbits for which the angular momentum of the whole *atom* is an integral multiple of $h/2\pi$" Thus

$$\mu r^2 \text{–} \omega = n \frac{h}{2\pi}. \qquad \text{... }(i)$$

In the absence of nuclear motion, the corresponding equation was

$$mvr = n \frac{h}{2\pi}.$$

But $v = r\omega$, so that

$$mr^2\omega = n \cdot \frac{h}{2\pi}. \qquad \text{...}(ii)$$

Eq. (*i*) is exactly the same as eq. (*ii*) except that μ has replaced *m*. Therefore all the results of Bohr theory (which assume mass of the nucleus infinite) still hold, provided *m* is replaced by μ.

The energy of the electron in the *n*th orbit is now

$$E = -\frac{\mu e^4}{8_0^{\epsilon 2} h^2}\left(\frac{1}{n^2}\right).$$

Since μ is slightly less than *m*, the electron energies are slightly less negative than if the nucleus were at rest (*i.e.* infinitely heavy). The wavelengths of the hydrogen lines are now given by

$$\frac{1}{\lambda} = \frac{\mu e^2}{8_0^{\epsilon 2} ch^3}\left(\frac{1}{n_{f^2}} - \frac{1}{n_i^2}\right)$$

$$= R_H \left(\frac{1}{n_f^2} - \frac{1}{n_i^2}\right),$$

where $R_H \left(\frac{\mu e^4}{8\epsilon_0^2 ch^3}\right)$ is the Rydberg constant *for the hydrogen atom*. Since the value of the Rydberg constant is slightly less than if the nucleus were at rest, *the wavelengths are slightly larger than those corresponding to an infinitely heavy nucleus.*

Effect of Finite Nuclear Mass on Rydberg Constant : Taking nuclear motion into account, the Rydberg constant *for the hydrogen atom* is

$$R_H = \frac{\mu e^4}{8\epsilon_0^2 ch^3}$$

Now $$\mu = \frac{M_H m}{M_H + m}.$$

$$\therefore \qquad R_H = \frac{me^4}{8\epsilon_0^2 ch^3}\left(\frac{M_H}{M_H + m}\right).$$

But $\frac{me^4}{8\epsilon_0^2 ch^3}$ is the value of the Rydberg constant of an infinitely heavy nucleus, and is a constant *for all the atoms.* Let it be denoted by R_∞. Its value is 109737 cm^{-1}. Thus

$$R_H = R\infty \left(\frac{M_H}{M_H + m}\right) = \frac{R\infty}{1 + (m/M_H)}.$$

This shows that the Rydberg constant for hydrogen, R_H depends upon the mass M_H of the hydrogen nucleus. Thus it is a constant only for a particular atom, and varies slightly from atom to atom. This variation, clearly, arises because of the finite mass (motion) of the nucleus. The Rydberg constant for hydrogen is $R_H = 109677$ cm^{-1}.

The variation of Rydberg constant from element to element is shown in figure below. The largest change occurs between H^1 and H^2. With increasing mass number, the Rydberg constant approaches more and more closely to R_∞.

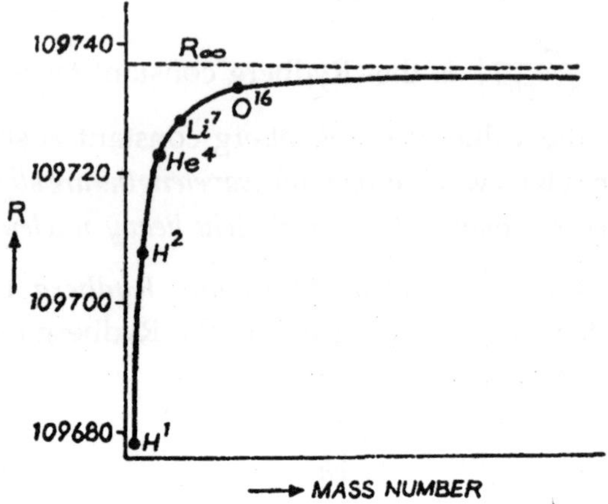

Heavy Hydrogen Discovered

The variation of Rydberg constant with the mass of the nucleus resulted in the discovery of deuterium (heavy hydrogen), The deuterium is an isotope of hydrogen whose nucleus has a mass almost exactly double that of ordinary hydrogen. Therefore, the Rydberg constant for deuterium is slightly greater than that for hydrogen. Consequently, the wavelengths of the spectral lines of deuterium are slightly shorter than those of the corresponding spectral lines of hydrogen. The general nature of the two spectra is, however, exactly same because atoms of both hydrogen and deuterium have the same electron structure.

Urey, in 1932, photographed hydrogen spectrum by means of a large concave grating. He observed that each hydrogen line was accompanied by a very faint line slightly on the short wavelength side. He measured the wavelength difference and attributed the faint line to some isotope of hydrogen. This isotope is now known as 'heavy hydrogen' or 'deuterium.

Bohr Theory Applied to Singly-ionised Helium Atom (He^+): The Bohr theory can be applied to ions containing single electron such as He^+ (singly-ionised helium), Li^{++} (doubly-ionised lithium), and so on. Such ions behave just like hydrogen atom except that they have greater nuclear charge and mass. Hence they are also called as 'hydrogen-like atoms'.

The nucleus of a hydrogen-like atom of atomic number Z carries a charge $+Ze$ so that the electrostatic force of attraction which it exerts on the orbital electron is $\frac{1}{4\pi\epsilon_0}\frac{Ze^2}{r^2}$, instead of $\frac{1}{4\pi\epsilon_0}\frac{e^2}{r^2}$ as in the case of hydrogen atom for which Z = 1. Consequently, the formula for the energy levels of a hydrogen-like atom is

$$E = -\frac{\mu Z^2 e^4}{8\epsilon_0^2 h^2}\left(\frac{1}{n^2}\right),$$

where μ is the reduced mass of the electron in the hydrogen-like atom, which is different than that in *H*. The corresponding wave-length JA are given by

$$\frac{1}{\lambda} = \frac{\mu Z^2 e^4}{8\epsilon_0^2 ch^3}\left(\frac{1}{n_f^2} - \frac{1}{n_i^2}\right),$$

Now for He^+, we have Z = 2 and $\frac{\mu e^4}{8\epsilon_0^2 ch^3} = R_H$, (the Rydberg constant for ionised helium), μ now being the reduced mass of the electron in He^+. Therefore

$$\frac{1}{\lambda} z= 4R_{He}\left(\frac{1}{n_f^2} - \frac{1}{n_i^2}\right),$$

The formula for hydrogen (Z = 1) is

$$\frac{1}{\lambda} = R_H\left(\frac{1}{n_f^2} - \frac{1}{n_i^2}\right),$$

Relationship between H and He⁺ Spectra : The formula for He^+ differs from the formula for hydrogen by a factor of 4, provided we ignore the difference between R_{He} and R_H. The corresponding helium wavelengths are therefore shorter by a factor of 4 . It is clear from the above two formulae that a transition from $n_i = 4$ to $n_f = 2$ in He$^+$ would emit a photon of the same wavelength as that emitted during a transition from $n_i = 2$ to $n_f = 1$ in H. Hence we can conclude that the He$^+$ spectral lines coincide with certain H spectral lines. The coincident is, however, not exact because the Rydberg constant, for H and He$^+$ are slightly different.

Below given figure shows Balmer series of H and the Pickering series of He$^+$ photographed on the same plate. We see that alternate members of Pickering series of He$^+$ coincide roughly with members of Balmer series of H.

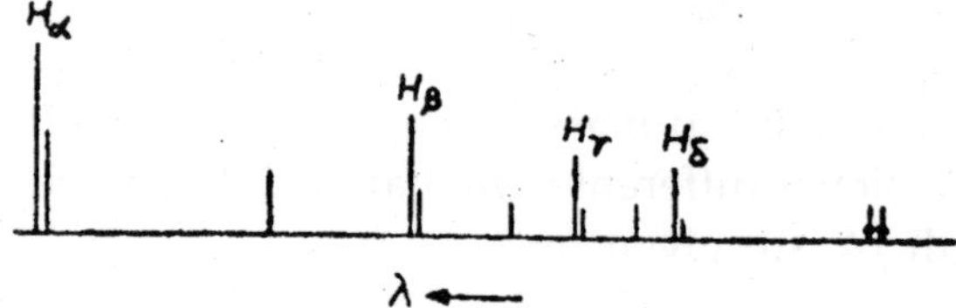

Determination of m/M$_H$:— The Rydberg constant for hydrogen atom is given by

$$R_H = \frac{R\infty}{1+(m/M_H)}, \qquad \text{... } (i)$$

where R_∞ is the Rydberg constant for an infinitely heavy nucleus, m is the mass of electron and M_H is the mass of hydrogen nucleus (*i.e.* mass of proton). Similarly, the Rydberg constant for He$^+$ is given by

$$R_{He} = \frac{R\infty}{1+(m/M_{He})},$$

where M_H, is the mass of a helium nucleus. Since helium nucleus is very nearly equal to four times hydrogen nucleus ($M_H \simeq 4M_H$), we write

$$R_{He} = \frac{R\infty}{1+(m/4M_H)}. \qquad \text{... } (ii)$$

Dividing eq. (*i*) by (*ii*), we get

$$\frac{R_H}{R_{He}} = \frac{1+(m/4M_H)}{1+(m/M_H)}$$

or $$R_H\left(1+\frac{m}{M_H}\right) = R_{He}\left(1+\frac{m}{4M_H}\right)$$

or $$\frac{m}{m_H}\left(R_H-\frac{1}{4}R_{He}\right) = R_{He}-R_H$$

or $$\frac{m}{M_H} = \frac{R_{He}-R_H}{R_H-\frac{R_{He}}{4}}.$$

R_{He} and R_H can be found by measurements of wavelengths in the series of helium and hydrogen spectra. Hence m/M_H, the ratio of the mass of electron to the mass of proton (*i.e.* hydrogen nucleus) can be determined.

Sommerfeld's Extension of Bohr Theory: According to Bohr's theory, an atom has a number of discrete energy levels (orbits), which only can be occupied by electrons. Each level is characterised by an integer *n* called the 'quantum number.' In a hydrogen atom, the (quantised) energy of a level with quantum number *n* is given by (ignoring nuclear motion)

$$En = -\frac{m\,e^4}{8\epsilon_0^2\,h^2}\left(\frac{1}{n^2}\right),$$

where *m*, *e* and *h* have their usual meanings. This we can write as

$$En = -\frac{R\,hc}{n^2} \qquad \text{... }(i)$$

where $R\left(=\frac{me^4}{8\epsilon_0^2\,ch^3}\right)$ is the Rydberg constant and *c* is the velocity of light. Whenever the electron jumps from one level to another, electromagnetic radiation of a *discrete* frequency is emitted (or absorbed). For example, when the electron jumps from the $n = 3$ to $n = 2$ level, the first line of Balmer series (called H_a-line) is emitted (see Fig.) and its frequency is

$$v = \frac{E_3 - E_2}{h}.$$

in a similar way, the emission of all spectral lines can be explained.

According to the above theory, each line of the hydrogen spectrum must be a *single* line *i.e.* it must have a single frequency (or wavelength). It was, however, observed by Michelson and others that in fact, the hydrogen lines were not strictly single. Each line consisted of a small number of close components of slightly different frequencies. This is called the 'fine structure' of spectral lines. This infers that each energy level of a given n must be splitted into sub-levels of slightly different energies. Hence the fine-structure could not be explained by Bohr's original theory,

Sommerfeld extended the Bohr's theory by assuming the existence of elliptical orbits (in addition to Bohr's circular orbits) for the electron, with one focus at the nucleus of the atom (Fig.). Now, an electron in an elliptic orbit has *two* degrees of freedom, namely the radial distance r and the azimuthal angle θ. Sommerfeld postulated that *each of these degrees of freedom must be quantised separately.* If $p\theta$ and p, be the angular and radial momenta of the electron, then according to Wilson Sommerfeld quantisation rules ; we must have

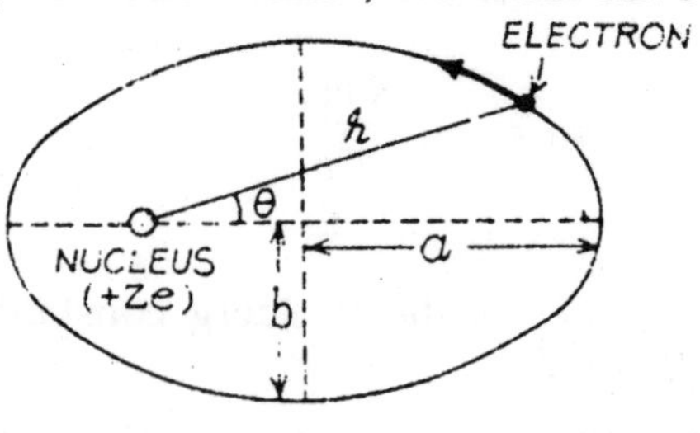

$$\int_0^{2x} pe\, d\theta = \text{kh}, \qquad \text{k} = 1, 2, 3, \dots \dots \qquad \dots (ii)$$

and $\oint p_r\, dr = n_r h,$ nr = 0, 1, 2, ... *(iii)*

where k and n_r are integers, called respectively 'azimuthal' and 'radial' quantum numbers. By integrating over a complete revolution, it can be proved that

$$1- e^2 = k^2 / K + n_r)^2$$

where e is the eccentricity of the ellipse. Since both k and n, ate integers, we can put

$$k + n_r = n, \qquad (n = 1, 2, 3,.....)^*$$

where n is called the 'total quantum number' of the electron. Thus

$$1 - \epsilon^2 = \frac{k^2}{n^2}.$$

If a and b be the semi-major and semi-minor axes of the ellipse, then $(1 - e^2) = b^2\ a^2$. Therefore

$$\frac{b^2}{a^2} = \frac{k^2}{n^2}$$

or

$$\boxed{\frac{b}{a} = \frac{k}{n}.} \qquad \text{... (iv)}$$

This is the condition of quantisation for the orbits. *Only those elliptic orbits are permitted for the electron for which the ratio of the major to the minor axis is the ratio of two integers.*

Let us now calculate the total energy E of an electron in a quantised elliptic orbit. The energy is the sum of the kinetic energy K and potential energy U, That is

$$E = K + U$$

$$= \frac{1}{2} m\, (r^2 + r^2\theta^2) - \frac{1}{4\pi_0^\epsilon} \frac{e^2}{r}.$$

But $p_r = mr$ and $p_\theta = mr^2\theta$. Therefore the last expression may be written as

$$E = \frac{1}{2m}\left\{p_r^2 + \frac{p_\theta^2}{r^2}\right\} - \frac{1}{4\pi\epsilon_0}\frac{e^2}{r}.$$

Now, $p_r = m\dot{r} = m\frac{dr}{dt} = m\frac{dr}{d\theta}\frac{d\theta}{dt} = m\frac{dr}{d\theta}\dot{\theta} = (mr^2\dot{\theta})\frac{1}{r^2}\frac{dr}{d\theta}$

$$= \frac{p_\theta}{\gamma^2}\frac{dr}{d\theta}$$

Putting this value of p_r in the last eq., we get

$$E = \frac{p_\theta^2}{2mr^2}\left[\left(\frac{1}{r}\frac{dr}{d\theta}\right)^2 + 1\right] - \frac{1}{4\pi\epsilon_0}\frac{e^2}{r}. \quad \text{... }(v)$$

Now, the polar equation of the ellipse is

$$\frac{1}{r} = \frac{1}{a}\frac{1-\epsilon\cos\theta}{1-\epsilon^2} \quad \text{... }(vi)$$

where $(1-\epsilon^2) = b^2/a^2$. If we put the value of r and $\frac{dr}{d\theta}$ from eq. (*vi*) in eq. (*v*), then we can show that

$$E = -\frac{me^4\left(1-\epsilon^2\right)}{\left(4\pi\epsilon_0\right)^2 2p_\theta^2}. \quad \text{... }(vii)$$

For an isolated system, the angular momentum p_θ is constant. Then from eq. (*ii*), we get

$$p_\theta = \frac{kh}{2\pi}.$$

Also $\quad (1-\epsilon^2) = \frac{b^2}{a^2} = \frac{k^2}{n^2}.$ [from eq. (*iv*)]

Making these two substitutes in eq, (vii), we get

$$E = -\frac{me^4\left(\frac{k^2}{n^2}\right)}{\left(4\pi\epsilon_0\right)^2 2\left(\frac{kh}{2\pi}\right)^2}$$

$$= -\frac{me4}{8\epsilon_0^2 h^2}\left(\frac{1}{n^2}\right).$$

Making the substitution, $R = \dfrac{me4}{8\epsilon_0^2 ch^3}$, we get

$$E = -\frac{Rhc}{n^2}. \qquad \text{... }(viii)$$

This is exactly the same as eq. (*i*). the energy of electron in a circular Bohr orbit.

Now, for a given value of n $(= k + n_r)$, k can take n possible different values ($k = 1, 2, 3, \ldots n$). This means that for a given n, there are n orbits of *different* eccentricities (according to the condition $\left(\frac{b}{a} = \frac{k}{n}\right)$ which can be occupied by the electron, But *the* energy E for all these orbits is the same, because E is independent of k and depends upon n only [see eq. (*viii*)]. Thus *the mere introduction of elliptic orbits adds no new energy levels.*

Role of Relativistic Correction

In an elliptic orbit the velocity of the electron varies, becoming very large when the electron passes close *to* the nucleus. The theory of relativity shows that when the velocity of a particle increases, its mass also increases. This effect makes the energy of the electron in a more elliptic *(i.e.* of greater eccentricity) orbit greater than that in a less elliptic orbit. Thus, *the different k-orbits with a given n have slightly different energies,* according to the following expression, also proved by Sommerfeld

$$E = \frac{Rhc}{n^2} - \frac{Rhc\,\alpha^2}{n^3}\left(\frac{1}{k} - \frac{3}{4n}\right),$$

where α is a pure number ($\simeq 1/137$) called the 'fine-structure constant'.

We can now explain the fine structure of H_α line which is emitted when the electron jumps from the $n = 3$ to $n=2$ level. The $n = 3$ level consists of 3 sub-levels corresponding to $k = 1$, 2, 3, and the $n=2$ level consists of 2 sub-levels corresponding to $k=1,2$ (Fig.). Only those electron-jumps from the upper sub-levels to the lower sub-levels are allowed for which k changes by unity ($\Delta k = \pm 1$). Thus there are three possible jumps, showing that the H_α*-line* actually consists of three components.

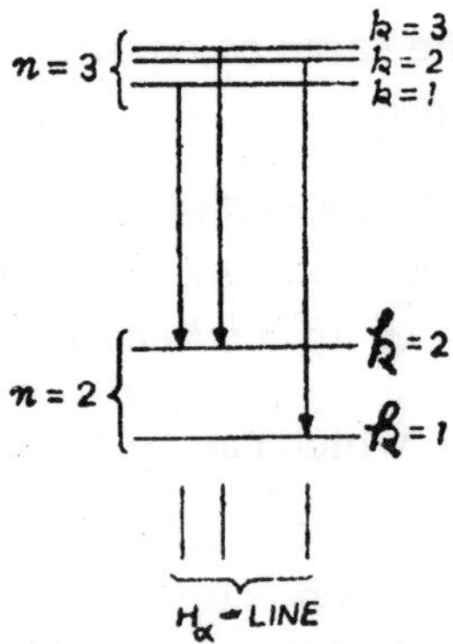

Note: From quantum mechanical considerations the azimuthal quantum number k has been replaced by $I + I$. Hence l can take the values 0, 1, 2, ... $(n - 1)$.

Excitation and Ionisation Potentials of an Atom: According to the Bohr's theory, in an atom there are certain discrete orbits only in which an electron can revolve without radiating energy. Each of these orbits of a given atom is characterised by a certain (quantised) energy. Hence we say that there are certain discrete energy levels in an atom.

When an electron in an atom absorbs sufficient energy from an outside source, it rises from its present energy level to a higher level. The atom is then said to be 'excited'. This excited state lasts only for about 10^{-8} seconds after which the electron jumps back to the inner level, emitting the absorbed energy in the form of electromagnetic radiation.

When the electron in the atom gains sufficient energy so as to escape *completely* from the atom, the atom is said to be 'ionised'.

Again when the atom recaptures an electron and reverts to its original state, the absorbed energy is emitted as electromagnetic radiation.

The energy required to excite or to ionise a given atom is perfectly definite. For example, in case of hydrogen atom the energy of the nth orbit is

$$E_n = -\frac{me^4}{8\epsilon_0^2 h^2}\left(\frac{1}{n^2}\right).$$

Substituting the known values of m (= 9.1×10^{-31}kg), e(= 1.6×10^{-19} coul), h(= 6.62×10^{-34} joule-sec) and ϵ_0 (= 8.85 $\times 10^{-20}$ coul^2/nt-m^2), we get

$$E_n = -2.17 \times 10^{-18}\left(\frac{1}{n^2}\right) \text{ joule}$$

$$= -\frac{2.17 \times 10^{-18}}{1.6 \times 10^{-19}}\left(\frac{1}{n^2}\right)$$

$$= -\frac{13.6}{n^2} \text{ electron-volt.}$$

[1eV = 1.6×10^{-19} joule]

Putting $n = 1, 2, 3, \ldots\ldots \infty$, the energies of the 1st, 2nd, 3rd,......∞ energy levels come out to be —13.6, —3.4, —1.51,......0 electron-volts respectively. Hence the energy to be supplied to the atom to raise an electron from the first to the second orbit is (13.6 – 3.4) = 10.2 electron-volts, from first to the third orbit is (13.6 – 1.51) = 12.09 electron-volts,* and to the ionized state is (13.6 – 0) = 13.6 electron-volts.

Now, the atoms are usually excited or ionised by bombarding them with electrons accelerated through a potential. But the excitation or unionisation of an atom can take place only when the bombarding electron has the required amount of energy.

The minimum accelerating potential which imparts to the bombarding electron an energy sufficient to excite a given atom is called the 'excitation potential' of the atom. Similarly, *the minimum accelerating potential which imparts to the bombarding electron an energy sufficient to ionise the atom is called the 'ionisation potential' of the atom.*

Thus 10.2, 12.09, electron-volts are the excitation potentials; while 13 6 electron-volts is the ionisation potential for the hydrogen atom. All these together are called the 'critical potentials' of the atom.

Experiment of Franck-Hertz

Franck and Hertz, in 1914, performed a series of experiments to measure the excitation potentials of atoms of different elements. These experiments showed directly that in an atom discrete energy levels do exist.

Their apparatus is shown in Figure below. It consists of a glass tube in which are mounted a filament *F*, a grid *G* , and a plate *P* as shown. The filament *F* is heated by a small battery *B*. An accelerating potential *V* is applied between *F* and *G* ; and a *small fixed* retarding potential V_0 (about 0.5 volt) between G and *P*. The gas of the element whose atoms are to be studied is introduced in the tube at about 1 mm of mercury pressure.

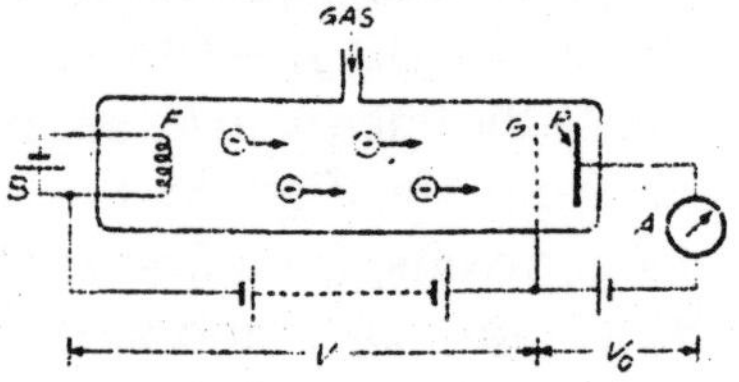

The electrons emitted from the hot filament *F* are accelerated between *F* and *G* by the potential *V*, and retarded between *G* and *P* by the potential V_0. Thus only electrons having energies greater than, eV_0 at *G* are able to reach *P*. The current to *P* is recorded by an ammeter *A*.

The current is plotted against the *accelerating* potential *V*. which is gradually increased from zero. The curve obtained shows a series of regularly-spaced peaks (Fig.).

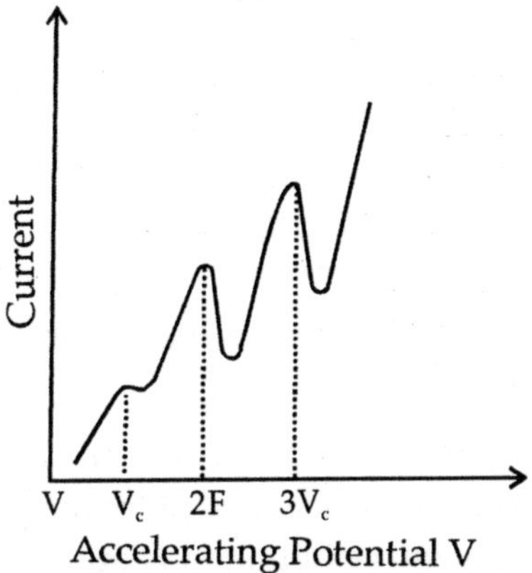

Interpretation of the Curve: Electrons are emitted from *F* with a range of small energies. In the beginning, they acquire a *small* additional energy e*V* on reaching *G*. Those electrons whose energy is now greater than eV_0 reach *P*, and a current is obtained. As the accelerating potential *V* is increased, more and more of the electrons arrive at the plate and the current rises. Between *F* and *G* the electrons collide with the gas atoms. But since the electron energies are not enough to excite the atoms, the electrons do not lose energy in these (elastic) collisions.

When the accelerating potential *V* becomes equal to the first excitation potential V_e of the gas atoms, the electron energy at G is cV_e. The electrons can now suffer *inelastic* collisions with gas atoms near *G* and excite them to an energy level above their ground state. The electrons which do so, lose their energy and are unable to reach P against the retarding potential. Hence the current drops sharply. Thus the position of the first peak gives the excitation potential V_c of the gas atoms.

As the accelerating potential V is increased further, the electrons suffer inelastic collisions nearer and nearer to the filament F, so that when they reach the grid G, once again they acquire enough energy to reach the plate P. The current thus begins to rise again. When V becomes equal to $2V_c$, a second inelastic collision occurs near the grid and the current drops again. Thus the second peak gives 2 V_c, . This process repeats as V is further increased.

The first peak occurs at a potential slightly less than V_c. This is because electrons are emitted from the filament with a finite velocity and hence with some energy. The true excitation potential V_e, is obtained by measuring the difference between two successive peaks.

Limitations : Atoms have more than one excitation potentials and also an unionisation potential. Therefore in an actual experiment the curve obtained is quite complicated and we cannot distinguish between which are excitation and which are unionisation potentials

Demonstration of the Existence of Discrete Energy Levels: The experiment shows that electrons transfer energy to the atoms in discrete amounts, and that they cannot excite atoms if their energy is less than *eVc*. Franck and Hertz demonstrated it directly by observing the spectrum of the gas during electron collisions. They showed that a particular spectral line does not appear until the electron energy reaches a *threshold* value. For example in the case of mercury vapour, they found that a minimum electron energy of 4.9 eV was required to obtain the 2536 A line of mercury; and a photon of 2536Å light has an energy of just 4.9 eV. This shows that discrete energy levels *do* exist in atom and the electrons of the atom can exist only in these levels.

Bohr's Correspondence Principle : The quantum theory gives results which are altogether different from those given by the classical theory in the microscopic world. Yet the two theories approach each other as the quantum number in question increases. This fact was pointed out by Bohr in 1923 who enunciated the following principle :

"The predictions of the quantum theory for the behaviour of any physical system must coincide with the predictions of the classical theory in the limit in which the quantum numbers specifying the state of the system become very large." This is known as 'Bohr's correspondence principle'. We establish it by means of an example.

According to classical theory, an electron moving in a circular orbit must emit radiation only of a particular frequency, namely, the frequency of revolution of the electron itself (or its integral multiples). According to quantum theory, on the other hand, radiation is emitted when the electron jumps from one orbit to the other and the frequency of radiation is determined by the difference in energy between the two orbits. Let us try to find a relation between the two frequencies.

The basic equations in the Bohr's theory of the hydrogen atom are the following :

$$\frac{mv^2}{r} = \frac{1}{4\pi \epsilon_0} \frac{e^2}{r^2}$$

and $$mvr = n\frac{h}{2\pi} \qquad n = 1, 2, 3, \ldots$$

where v is the velocity of the electron of mass m in the orbit of radius r. These equations give

$$v = \frac{nh}{2\pi mr}$$

and $$r = \frac{n^2 h^2 \epsilon_0}{\pi m e^2}.$$

Hence the classical frequency of revolution of the electron is

$$f = \frac{\text{electron speed}}{\text{orbit circumference}} = \frac{v}{2\pi r}$$

$$= \frac{nh}{4\pi^2 mr^2} = \frac{nh}{4\pi^2 m\left(\dfrac{n^2h^2 \epsilon_0}{\pi me^e}\right)^2}$$

$$= \frac{me^4}{4 \epsilon_0^2 n^3h^3} = \frac{me^4}{8 \epsilon_0^2 ch^3}\frac{2c}{n^3}$$

$$= \frac{2Rc}{n^3}, \qquad \ldots (i)$$

where $R \left(= \dfrac{me^4}{8 \epsilon_0^2 ch^3}\right)$ is the Rydberg constant. This is the frequency which must be radiated by the moving electron classically.

Now, the frequency radiated on the basis of quantum theory is given by

$$v = \frac{E_i - E_f}{h}$$

$$= \frac{me^4}{8 \epsilon_0^2 h^3}\left(\frac{1}{n_f^2} - \frac{1}{n_i^2}\right),$$

when the electron drops from the n_ith orbit to the n_fth orbit. Again, introducing R, we have

$$v = Rc\left(\frac{1}{n_f^2} - \frac{1}{n_i^2}\right).$$

This may be written in the form

$$v = Rc\frac{(n_i - n_f)(n_i + n_f)}{n_f^2 n_i^2} \qquad \ldots (ii)$$

Let us write $n_f = n$ and $n_i = n + 1$. Then the frequency of the emitted radiation for the transition $n + 1 \rightarrow n$ (so that $\Delta n = 1$) is given by

$$v = Rc \frac{2n+1}{n^2 (n+1)^2}.$$

When n is very large, we can write $\frac{2n+1}{n^2 (n+1)^2} \simeq \frac{2}{n^3}$.

Under this condition, the emitted frequency is

$$v = \frac{2Rc}{n^3} \qquad ...(iii)$$

Equations (*i*) and (*iii*) yield

$$v = f$$

If we consider transitions $\Delta n = 2, 3, 4,$......we shall have

$$v = 2f, 3f, 4f, ...$$

Thus for very large quantum numbers *n*, the quantum-theory frequency v of the radiation is identical with the classical frequency of the revolution (or its harmonics) of the electron in the orbits. This is in accordance with Bohr's correspondence principle.

PROBLEMS

1. ***The speed of electron in the nth orbit of the hydrogen atom is calculated. If relativist correction is important for*** $\frac{v}{c} > 0.005$, ***find for which orbit this correction is necessary, (e = 16 ×10***^{19} ***coul, h = 6.63 ×*** 10^{-34} ***joule-see, c = 3.0 ×*** 10^8 ***m/sec and*** $\frac{1}{4\pi\epsilon_0} = 9 \times 10^9$ ***nt-m***^{2}***/coul***^{2}***).***

Solution: Let v be the speed of the electron (mass *m*, charge *e*) *in* the *n*th orbit whose radius is *r*, all in M.K.S. units. The charge on hydrogen nucleus is + e. As the centripetal force is provided by the electrostatic attraction, we have

$$\frac{mv^2}{r} = \frac{1}{4\pi\epsilon_0}\frac{e^2}{r^2}. \qquad \text{... }(i)$$

Further, from Bohr's postulate, the angular momentum of the electron is an integral multiple of $h/2\pi$, that is,

$$mvr = \frac{nh}{2\pi}, \qquad n = 1, 2, 3, \ldots \ldots\ldots (ii)$$

Dividing eq. (*i*) by eq. (*ii*), we get

$$v = \frac{1}{4\pi\epsilon_0}\frac{2\pi e^2}{nh}.$$

Substituting the given values, we get

$$v = (9 \times 10^9 \text{ nt-m}^2/\text{coul}^2) \times \frac{2 \times 3.14 \times (1.6 \times 10^{-19} \text{ coul})^2}{n(6.63 \times 10^{-34} \text{ joule-sec})}$$

$$= \frac{1}{n}(2.18 \times 10^6) \text{ m/sec.}$$

This is the required value. Now

$$\frac{v}{c} = \frac{1}{n}\frac{2.18 \times 10^6}{3.0 \times 10^8} = \frac{0.0073}{n}.$$

Thus, for $n = 1, 2, 3, \ldots \ldots$ we have

$$\frac{v}{c} = 0.0073, 0.0036, 0.0024, \ldots \ldots$$

Hence the relativist correction is necessary for $n = 1$ orbit only.

2. ***An expression for the radius of the electron orbit in the hydrogen atom in its nth state is found in the below-given way. What will be the approximate quantum number n for an electron in an orbit of radius 1 mm is also given below (Take required values from problem 1).***

Solution: The basic equations in the Bohr's theory of hydrogen atom are :

$$\frac{mv^2}{r} = \frac{1}{4\pi\epsilon_0}\frac{e^2}{r^2} \qquad \text{... } (i)$$

and $$mvr = \frac{nh}{2\pi}. \qquad \text{... } (ii)$$

Squaring (ii) and dividing by (i). we get

$$r = 4\pi\epsilon_0 \frac{n^2h^2}{4\pi^2 me^2}.$$

Substituting the given values, we get

$$r = \frac{1}{9 \times 10^9 \text{ nt-m}^2/\text{coul}^2}$$

$$\times \frac{n^2\left(6.63 \times 10^{-34} \text{ joule-sec}\right)^2}{4 \times (3.14)^2 \times \left(9.1 \times 10^{-31} \text{ kg}\right) \times \left(1.6 \times 10^{-19} \text{ coul}\right)^2}$$

$$= 0.53 \times 10^{-10}\, n^2 \text{ meter}$$

$$= 0.53\, n^2 \text{ Å}.$$

For $r = 1 \text{ mm} = 10^7$ Å, we have

$$10^7 = 0.53\, n^2$$

$$\therefore \quad n^2 = \frac{10^7}{0.53} = 18.87 \times 10^6$$

or $$n \simeq 4350.$$

3. *The time taken by the electron to traverse the first orbit in the hydrogen atom is calculated as follows. Electron mass and charge are 91 x 10-^{31} kg and l.6 ×10–^{19} coul and h = 6.63 × 10^{34} J-s.*

Solution: The radius of the *n*th Bohr orbit is

$$r_n = 4\pi\epsilon_0 \frac{n^2h^2}{4\pi^2 me^2}$$

and the velocity of electron in the nth orbit is

$$v_n = \frac{1}{4\pi\epsilon_0}\frac{2\pi e^2}{nh}.$$

The time by the electron to traverse the nth orbit is therefore

$$T_n = \frac{2\pi\, e\, r_n}{v_n}$$

$$= 2\pi\left(4\pi\epsilon_0\right)^2 \frac{n^2h^2}{4\pi^2 me^2}\,\frac{nh}{2\pi e^2}$$

$$= \frac{4\epsilon_0^2 h^3 n^3}{me^4}.$$

For the first orbit, $n = 1$. Substituting the given values of h, m, e and $\epsilon_0 = 8.85 \times 10^{-12}\ C^2/\ \text{N-m}^2$, we get

$$T_1 = \frac{4\left(8.85\times 10^{-12}\right)^2\left(6.63\times 10^{-34}\right)^3}{\left(9.1\times 10^{-31}\right)\left(1.6\times 10^{-19}\right)^4}$$

$$= 1.5 \times 10^{-16}\ \text{s}.$$

4. *The average life-time of an electron in an excited state of hydrogen atom is about 10^{-8} sec. Let us find as to how many revolutions does an electron in the $n = 2$ state make before dropping to the $n = 1$ state ($R = 1.097 \times 10^7$ meter^{-1}).*

Solution: Let v be the velocity of electron (mass m, charge e) in an orbit of radius r. The basic equations are

$$\frac{mv^2}{r} = \frac{1}{4\pi\epsilon_0}\frac{e^2}{r^2}$$

and

$$= mvr = n\frac{h}{2\pi}.$$

These equations give

$$v = \frac{nh}{2\pi mr}, \ r = \frac{n^2h^2 \in_o}{\pi me^2}.$$

The number of revolutions of the electron in the orbit per second is

$$f = \frac{v}{2\pi r} = \frac{2Rc}{n^3} \qquad \text{(as proved in Q, 8)}$$

For $n = 2$ state, we have

$$f = \frac{2 \times \left(1.097 \times 10^7 \text{ meter}^{-1}\right) \times \left(3 \times 10^8 \text{ meter / sec}\right)}{8}$$

$$= 8.2 \times 10^{14} \text{ per sec.}$$

Hence the number of revolutions of the electron in its life-time of 10^{-8} second.

$$= (8.2 \times 10^{14}) \times 10^{-8} = \mathbf{8.2 \times 10^6}.$$

5. ***The wavelength of the photon emitted when the hydrogen atom goes from n = 10 state to the ground state is found as follows.*** **(R_H= 1.097 × 10^{-3} A^{-1}).**

Solution: Since the atom drops to the ground state (n = 1), the photon emitted belongs to the Lyman series. The wavelengths of the spectral lines in this series are given by

$$\frac{1}{\lambda} = R_H\left(\frac{1}{1^2} - \frac{1}{n^2}\right); \qquad n = 2, 3, 4, \ldots\ldots$$

For $n = 10$, we have

$$\frac{1}{\lambda} = R_H\left(1 - \frac{1}{100}\right) = \frac{99}{100} R_H$$

or

$$\lambda = \frac{100}{99 R_H}$$

$$= \frac{100}{99 \times \left(1.097 \times 10^{-3} \overset{\circ}{A}^{-1}\right)} = 921 \text{ Å}$$

6. The wavelength of the first line of Balmer series is 6563 A. Rydberg constant is calculated follows.

Solution: The wavelengths of the spectral lines of Balmer series are given by

$$\frac{1}{\lambda} = R_H \left(\frac{1}{2^2} - \frac{1}{n^2}\right); \quad n = 3, 4, 5, \dots \dots$$

For the first line ($\lambda = 6563 \times 10^{-10}$ m), $n = 3$.

$$\therefore \quad \frac{1}{6563 \times 10^{-10} \text{ m}} = R_H \left(\frac{1}{2^2} - \frac{1}{3^2}\right) = \frac{5}{36} R_H$$

or

$$R_H = \frac{36}{5 \times \left(65 \times 10^{-10} \text{ m}\right)} = 1.097 \times 10^7 \text{ m}^{-1}.$$

7. The wavelengths of the first two lines of the Balmer series and the series limit is calculated as follows. (h = 6.63 ×10^{-34} joule-sec, c = 3.0 × 10^8 meter/sec, e = 1.6 × 10^{-19} coul, m — 9.1 × 10^{-31} kg, ϵ_0 = 8.85 × 10^{-12} coul^2/nt-m^2).

Solution: Let us first calculate the Rydberg constant from the given data. We have

$$R = \frac{me^4}{8 \epsilon_0^2 ch^3}$$

$$= \frac{\left(9.1 \times 10^{31} \text{ kg}\right) \times \left(1.6 \times 10^{-19} \text{ coul}\right)^4}{8 \times \left(8.85 \times 10^{12} \text{ coul}^2 / \text{nt} - \text{m}^2\right)^2 \times \left(3.0 \times 10^8 \text{ m/sec}\right) \times \left(6.63 \times 10^{-34} \text{ joule-sec}\right)^2}$$

$= 1.097 \times 10^7$ meter-^{1}.

Now, the wavelengths of the spectral lines of Balmer series are given by

$$\frac{1}{\lambda} = R\left(\frac{1}{2^2}-\frac{1}{n^2}\right); \qquad n = 3, 4, 5, \ldots\ldots$$

For the first line, $n = 3$.

$$\therefore \qquad \frac{1}{\lambda} = R\left(\frac{1}{2^2}-\frac{1}{3^2}\right) = \frac{5R}{36}$$

$$\text{or } \lambda_1 = \frac{36}{5R} = \frac{36}{5\times\left(1.097\times10^7 \text{ meter}^{-1}\right)} = 6.563 \times 10^{-7} \text{ m}$$

$$= 6563 \times 10^{-10} \text{ meter} = 6563 \text{ Å}.$$

For the second line,$n = 4$. We can see that $\lambda_2 = 4861$ Å.

For the series limit, n — ∞.

$$\therefore \qquad \frac{1}{\lambda\infty} = \frac{R}{4}$$

$$\text{or} \qquad \lambda\infty = \frac{4}{R} = \frac{4}{1.097\times10^7}$$

$$= 3.646 \times 10^{-7} \text{ meter} = 3646 \text{ Å}.$$

8. *The first member of Balmer series of hydrogen has a wavelength of 6563 Å. Calculate the wavelength of the second member is calculated as follows.*

Solution: The wavelengths of the spectral lines of Balmer series are given by

$$\frac{1}{\lambda} = \text{R}\left(\frac{1}{2^2}-\frac{1}{n^2}\right); \qquad n = 3, 4, 5 \ldots\ldots$$

For the first member, $n = 3$.

$$\therefore \qquad \frac{1}{\lambda_1} = R\left(\frac{1}{2^2}-\frac{1}{3^2}\right) = \frac{5R}{36} \qquad \ldots (i)$$

For the second member, $n = 4$.

$$\therefore \qquad \frac{1}{\lambda_2} = R\left(\frac{1}{2^2} - \frac{1}{4^2}\right) = \frac{3R}{16} \qquad \text{... (ii)}$$

Dividing eq. (*ii*) by (*i*), we get

$$\frac{\lambda_1}{\lambda_2} = \frac{3R}{16} \times \frac{36}{5R} = \frac{27}{20}.$$

$$\therefore \qquad \lambda_2 = \lambda_1 \times \frac{20}{27}$$

But $\lambda_1 = 6563$ Å (given).

$$\therefore \qquad \lambda_2 = (6563 \text{ Å}) \times \frac{20}{27} = 4861 \text{ Å}.$$

9. ***The first line of the Balmer series in the spectrum of hydrogen has a wavelength of 6563 Å. The wavelength of the first line of Lyman series in the same spectrum is calculated as follows.***

Solution: The wavelength of the spectral lines of hydrogen spectrum are given by

$$\frac{1}{\lambda} = R\left(\frac{1}{n_f^2} - \frac{1}{n_i^2}\right),$$

where R is Rydberg constant. For the first member of Balmer series, $n_i = 2$ and nt = 3.

$$\therefore \qquad \frac{1}{\lambda_1} = R\left(\frac{1}{2^2} - \frac{1}{3^2}\right) = \frac{5R}{36}. \qquad \text{... (i)}$$

For the first member of the Lyman series, $nf = 1$ and $n_i = 2$.

$$\therefore \qquad \frac{1}{\lambda_1'} = R\left(\frac{1}{1^2} - \frac{1}{2^2}\right) = \frac{3R}{4}. \qquad \text{... (ii)}$$

From (*i*) and (*ii*), we have

$$\frac{\lambda_1'}{\lambda_1} = \frac{5R}{36} \times \frac{4}{3R} = \frac{5}{27}.$$

$$\therefore \qquad \lambda_1' = \frac{5}{27}\lambda_1 = \frac{5}{27} \times 6563 = 1215 \text{ Å}.$$

10. ***The series limit of Balmer series is at 3646 Å. The wavelength of the first member of this series is calculated as follows.***

Solution: The formula for Balmer series is

$$\frac{1}{\lambda} = R\left(\frac{1}{2^2} - \frac{1}{n^2}\right), \qquad n = 3, 4, 5, \ldots$$

For series limit $n = \infty$ and $\lambda = \lambda_\infty, = 3646$ Å.

$$\therefore \qquad \frac{1}{3646 \text{ Å}} = \frac{R}{4}$$

$$\text{or} \qquad R = \frac{4}{3646 \text{ Å}}.$$

Again, for the first member $n = 3$.

$$\therefore \qquad \frac{1}{\lambda_1} = R\left(\frac{1}{2^2} - \frac{1}{3^2}\right) = \frac{5R}{36}$$

$$\text{or} \qquad \lambda_1 = \frac{36}{5R} = \frac{36}{5 \times \frac{4}{3646 \text{ Å}}} = 6563 \text{ Å}.$$

11. ***An orange photon of wavelength 600 nm is emitted from an atom. The difference in energy in the two atomic states involved and the same result for the red 6563 Å line of hydrogen are found out as follows. ($h = 6.63 \times 10^{-34}$ joule-sec $c = 3.0 \times 10^8$ meter/sec, 1 eV = 1.6 $\times 10^{-19}$ joule).***

Solution: By Bohr's postulate, the emitted frequency is given by

$$v = \frac{\Delta E}{h},$$

where ΔE is the difference in energy. But $v = \frac{c}{\lambda}$ Therefore

$$\frac{c}{\lambda} = \frac{\Delta E}{h}$$

or $$\Delta E = \frac{hc}{\lambda}.$$

Here $$\lambda = 600 \text{ nm} = 600 \times 10^{-9} \text{ meter}.$$

$$\therefore \quad \Delta E = \frac{(6.63 \times 10^{-34} \text{ joule-sec}) \times (3.0 \times 10^{2} \text{ meter / sec})}{600 \times 10^{-9} \text{ meter}}$$

$$= 3.31 \times 10^{-19} \text{ joule}$$

$$= \frac{3.31 \times 10^{-19}}{1.6 \times 10^{-19}} = 2.07 \text{ eV}.$$

For λ = 6563 Å = 6563 × 10^{-10} meter, we can show that

$$\Delta E = 3.03 \times 10\text{–}^{19} \text{ joule} = 1.9 \text{ eV}.$$

12. *The volume of energy required to remove an electron from a hydrogen atom in the ground state and also in a state with n = 8 is given as follows (ii) The corresponding energies of the singly-ionised helium atom is calculated. (m = 9.11 × 10^{-31} kg, e = 1.6 ×10^{-19} coul, h = 6.63 ×10^{-34} joule-sec, ϵ_0 = 8.85 × 10^{-12} coul^2/nt-m^2 and 1 eV = 1.6 × 10^{-19} joule).*

Solution: (*i*) The energy required to remove an electron from the hydrogen atom in the ground state (n = 1) to infinity (where the energy is zero) is numerically equal to the energy of the electron in the n = 1 orbit. Now, the energy of the electron in nth orbit of hydrogen (Z = 1) is given by

$$E_n = \frac{me^4}{8 \epsilon_0^2 h^2}\left(\frac{1}{n^2}\right)$$

For the ground orbit, n = 1.

$$\therefore \qquad E_1 = -\frac{me^4}{8\epsilon_0^2 h^2}$$

$$= -\frac{\left(9.11\times10^{31}\right)\times\left(1.6\times10^{-19}\right)^4}{8\times\left(8.85\times10^{-12}\right)^2\times\left(6.63\times10^{-34}\right)^2}$$

$$= -2.17 \times 10^{-18} \text{ joule}$$

$$= -\frac{2.17\times10^{-18}}{1.6\times10^{-19}} = -13.6 \text{ eV.}$$

Hence the energy required to remove the electron from $n = 1$ orbit to infinity is 13.6 eV.

This is known as the binding energy of the hydrogen atom.

Again, for $n = 8$ orbit,

$$E_8 = -\frac{me^4}{8\epsilon_0^2 h^2}\left(\frac{1}{8^2}\right) = -\frac{E_1}{64}$$

$$= -\frac{13.6}{64} = -0.213 \text{ eV.}$$

Hence the energy required to remove an electron from » = 8 orbit to infinity is 0.213 eV.

(*ii*) The energy expression for He^+ (Z = 2) is

$$(E_n)_{He+} = -\frac{mZ^2e^4}{8\epsilon_0^2 h^2}\left(\frac{1}{n^2}\right)$$

$$= -\frac{4me^4}{8\epsilon_0^2 h^2}\left(\frac{1}{n^2}\right) = -4\,(E_n)_H.$$

That is, the energies for He^+ are 4 times the corresponding energies for *H*. Thus the results are

4 × 13.6 =54.4 eV

4 × 0.213 =0.85 eV.

13. *The ground state energy of an electron in the hydrogen atom is —13.6 eV. The energy of (i) n = 3 state in He⁺, (ii) n = 2 state in Li⁺⁺ is computed as follows.*

Solution: The energy of an electron in the nth state of a hydrogen-like atom is given by

$$E_n = -\frac{mZ^2e^4}{8E_0^2h^2}\left(\frac{1}{n^2}\right)$$

$$= -\frac{RZ^2hc}{n^2}. \qquad \left[\therefore R = \frac{me^4}{8\epsilon_0^2 ch^3}\right]$$

For H, $Z = 1$, so that

$$(E_n)\,H = -\frac{Rhc}{n^2}.$$

When $n = 1$ (ground state), $(E_1)_H = -13.6$ eV (given).

$$\therefore \qquad Rhc = 13.6 \text{ eV}.$$

$$\text{Hence} \quad (E_2)_H = \frac{Rhc}{4} = -\frac{13.6}{4} = -3.4 \text{ eV}$$

$$\text{and} \quad (E_3)_H = \frac{Rhc}{9} = -\frac{13.6}{9} = -1.5 \text{ eV}.$$

(*i*) For He^+, $Z = 2$, so that

$$(E_3)_{He} = -\frac{4Rch}{9} = 4\,(E_3)_H = -6.0 \text{ eV}.$$

(*ii*) For Li^{++}, $Z = 3$, so that

$$(E_2)_{Li} = -\frac{9Rch}{4} = -30.6 \text{ eV}.$$

14. *The energy required to ionise hydrogen atom in the ground state, limit of Balmer series are calculated as follows. (Rydberg constant is 1.097 × 10⁷ m⁻¹, h = 6.63 × 10⁻³⁴ joule-sec, c = 3.0 × 10⁸ m/sec, 1 eV = 1.6 × 10⁻¹⁹ joule).*

Solution: The energy required to ionise *i.e.* to remove an electron from the hydrogen atom in the ground state ($n = 1$)

to infinity (where the energy is zero) is numerically equal to the energy of the electron in the n = 1 orbit. Now, the energy of the electron in nth orbit of hydrogen is given by

$$E_n = -\frac{me^4}{8\epsilon_0^2 h^2}\left(\frac{1}{n^2}\right)$$

$$= -\frac{Rhc}{n^2}. \qquad \left[\therefore R = \frac{me^4}{8\epsilon_0^2 ch^2}\right]$$

For the the ground (first) orbit, n = 1.

$$\therefore \quad E_1 = -Rhc$$

$$= -(1.097 \times 10^7 \text{ m}^{-1})(6.63 \times 10^{-34} \text{ J-s})(3 \times 10^8 \text{ m-s}^{-1})$$

$$= -21.8 \times 10^{-19} \text{ joule}$$

$$= -\frac{21.8 \times 10^{-19}}{1.6 \times 10^{-19}} = -13.6 \text{ eV}.$$

Hence the energy required to remove the electron from n = 1 orbit to infinity is 13.6 eV.

(*ii*) The Balmer series is represented by

$$\frac{1}{\lambda} = R\left(\frac{1}{2^2} - \frac{1}{n^2}\right); \qquad n = 3, 4, 5, \ldots\ldots$$

For the series limit $n = \infty$ so that

$$\frac{1}{\lambda} = \frac{R}{4}.$$

$$\therefore \quad \lambda = \frac{4}{R} = \frac{4}{1.097 \times 10^7 \text{ m}^{-1}}$$

$$= 3.646 \times 10^{-7} \text{ m} = 3646 \text{ Å}.$$

15. ***The wavelength of the first line of Balmer series of hydrogen is 6562.8 Å The ionisation potential and the first***

excitation potential of the hydrogen atom are calculated as follows. ($h = 6.63 \times 10^{-34}$ J-s, $c = 3 \times 10^8$ m/sec).

Solution: The wavelengths of the Balmer lines are given by

$$\frac{1}{\lambda} = R\left(\frac{1}{2^2} - \frac{1}{n^2}\right); \qquad n = 3, 4, 5, ...$$

For the first line $n = 3$ and $\lambda = 6562.8$ Å $= 6562.8 \times 10^{-10}$ m.

$$\therefore \quad \frac{1}{6562.8 \times 10^{-10}} = R\left(\frac{1}{2^2} - \frac{1}{3^2}\right) = \frac{5R}{36}$$

Or

$$R = \frac{36}{5 \times 6562.8 \times 10^{-10} \text{ m}}$$

$$= 1.097 \times 10^7 \text{ m}^{-1}$$

The ionisation potential of an atom is numerically equal to the (ionisation) energy required to remove an electron completely from the atom in the normal state. In hydrogen atom; the electron stays in the first orbit ($n = 1$). Hence the energy required to remove this electron to infinity (where the energy is considered to be zero) is numerically equal to the energy of the electron in the first orbit. We know that electron energy in hydrogen atom is given by

$$E_n = -\frac{Rhc}{n^2}$$

In the ground state, $n = 1$.

$$\therefore \quad E_1 = -Rhc$$

$$= -13.6 \text{ eV. (as obtained in the last problem).}$$

Hence the ionisation potential of the hydrogen atom is 13.6 volts.

(*ii*) The first excitation potential of the atom is the energy required to shift the electron from $n = 1$ to Now,

$$E_1 = -13.6 \textbf{ eV}$$

and $$E_2 = \frac{E_1}{4} = -3.4 \text{ eV}. \qquad \left[\therefore E_n \propto \frac{1}{n^2}\right]$$

$$\therefore \quad E_1 \sim E_2 = 13.6 - 3.4 = 10.2 \text{ eV}.$$

Hence the first excitation potential is 10.2 volts.

16. *A beam of electrons bombards a sample of hydrogen. Let us find as to through what minimum potential difference must the electrons have been accelerated if the first line of the Balmar series is to be emitted. The ionisation potential of hydrogen atom is 13.6 volts. The numbers of possible spectral lines can be expected if the atom finally attains the normal state is explained below.*

Solution: Normally the hydrogen atoms are in the ground state ($n = 1$). The first Balmar line is emitted when the atom returns from $n = 3$ to $n = 2$ state. Hence in order to emit this line the atoms must be first raised to the $n = 3$ state by electron bombardment. Therefore the energy of the bombarding electrons must be equal to the difference of energy between $n - 1$ and $n = 3$ states ($E_1 \sim E_3$).

The energy of hydrogen atom in the *nth* state is

$$E_n = -\frac{Rhc}{n^2}.$$

If E_1 is the energy in the ground ($n = 1$) state, then $E_1 = -Rhc$.

Thus $$E_n = \frac{E_1}{n^2}.$$

For $n = 3$;we have

$$E_3 = \frac{E_1}{9}.$$

$$\therefore \quad E_1 - E_3 = E_1 - \frac{E_1}{9} = \frac{8}{9} E_1.$$

But $E_1 = -13.6$ eV

$$\therefore \qquad E_1 - E_3 = \frac{8}{9} \times (13.6) = 12.1 \text{ eV}.$$

Hence the bombarding electrons must be accelerated by 12'1 volts.

If the atom finally attains the normal state, the possible number of spectral lines is 3 (3→2, 3→1, 2→1).

17. *A photon of energy 12.1 eV is completely absorbed by a hydrogen atom, originally in its ground state, so that the atom is excited. What is the quantum number of this state? The ground state (negative) energy of hydrogen atom is 13.6 eV.*

Solution : Let the required quantum number be n.

Then $$E_1 - E_x = -12.1 \text{ eV}$$

$$E_1 - \frac{E_1}{n^2} = -12.1 \text{ eV}. \qquad \left[\because E_n = \frac{E_1}{n^2}\right]$$

But $$E_1 = -13.6 \text{ eV}.$$

$$\therefore \qquad 1 - \frac{1}{n^2} = \frac{-12.1}{-13.6} = 0.89$$

or $$\frac{1}{n^2} = 1 - 0.89 = 0.11$$

or $$n^2 = \frac{1}{0.11} = 9$$

$$\therefore \qquad n = 3.$$

18. *A beam of monochromatic photons of energy 9 eV is incident on hydrogen gas all of whose atoms are in the ground state. It is found that the beam is fully transmitted without absorption. The reason is given ad follows. The ground state energy of an electron in the hydrogen atom is $E_1 = -13.6$ eV.*

Solution: The minimum energy that can be absorbed by ground-state hydrogen atom is $E_1 \sim E_2$, which would excite it to the next state (n = 2). Now,

$$E_1 \sim E_2 = E_1 \sim \frac{E_1}{4} \qquad \left[\therefore E_n = \frac{E_1}{n^2}\right]$$

$$= \frac{3}{4}E_1 = \frac{3}{4} \times 13.6 = 10.2 \text{ eV}.$$

Hence photons of energy 9 eV cannot be absorbed by hydrogen atoms.

19. ***The first two excitation potentials of atomic hydrogen in Franck-Hertz experiment are 10.2 and 12.09 volts. An energy level diagram and all possible transitions for emission and absorption along their wavelengths are shown as follows. (h = 6.63 ×10⁻³⁴ joule-sec, c = 3.0 × 10⁸. m/sec, 1 eV = 1.6 × 10⁻¹⁹ joule).***

Solution: The energy level diagram, and the *three* possible transitions (*a*), (*b*) and (*c*) for emission are shown in the figure.

The frequency of radiation resulting from the transition (a) is given by

$$v_\alpha = \frac{E_2 - E_1}{h}$$

and the corresponding wave length is

$$\lambda_\alpha = \frac{hc}{(E_2 - E_1)}. \qquad [\therefore c = v\lambda]$$

Now, from the Fig., $E_2 - E_1 = 10.2$ eV $= 10.2 \times (1.6 \times 10^{-19})$ joule. Therefore

$$\lambda_a = \frac{(6.63 \times 10^{-34} \text{ joule-sec}) \times (3.0 \times 10^8 \text{ m/sec})}{(10.2 \times 1.6 \times 10^{-19} \text{ joule})}$$

$$= 1.216 \times 10^{-7} \text{ m} = 1216 \times 10^{-10} \text{ m} = 1216 \text{ Å}.$$

Similarly, $\lambda_b = \dfrac{hc}{(E_3 - E_1)}$

$$= \frac{\left(6.63 \times 10^{-34}\right) \times \left(3.0 \times 10^{8}\right)}{12.09 \times 1.6 \times 10^{-19}}$$

$$= 1.026 \times 10^{-7}\,\text{m} = 1026 \times 10^{-10}\,\text{m} = 1026\ \text{Å}.$$

Also $$\lambda_c = \frac{hc}{\left(E_3 - E_2\right)} = \frac{hc}{\left(E_3 - E_1\right) - \left(E_2 - E_1\right)}$$

$$= \frac{hc}{\left(12.09 - 10.2\right)\text{eV}} = \frac{hc}{1.89\ \text{eV}}$$

$$= \frac{\left(6.63 \times 10^{-34}\right) \times \left(3.0 \times 10^{8}\right)}{\left(1.89 \times 1.6 \times 10^{-19}\right)}$$

$$= 6.567 \times 10^{-7}\,\text{m} = 6567 \times 10^{-10}\,\text{m} = 6567\ \text{Å}.$$

In absorption, only the transitions starting from $n = 1$ shall be observed which correspond to 1216 Å and 1026 Å.

20. *With Franck-Hertz type of experiment on sodium, the first spectral line to appear is the D-line, $\lambda = 5.89 \times 10^{-7}$ m. The first excitation potential of sodium is given below. Given : $h = 6.63 \times 10^{-34}$ joule-sec, $c = 3 \times 10^{8}$ m/sec and leV = 1.6×10^{-19} joule.*

Solution: Let V volt be the excitation potential. Then the (excitation) energy imparted to the electron will be eV joule, where e coulomb is the charge on the electron. This energy is re-emitted as photon (radiation) when the electron returns to the normal state. If v be the frequency of the emitted radiation, the photon energy will be hv, Thus

$$eV = hv$$

or $$V = \frac{hv}{e}$$

If λ be the wavelength of the emitted radiation, then $v = c/\lambda$.

$\therefore$ $$V = \frac{hv}{e\lambda}$$

$$= \frac{\left(6.63 \times 10^{-34} \text{ J-s}\right)\left(3 \times 10^{8} \text{ ms}^{-1}\right)}{\left(1.6 \times 10^{-19} \text{ C}\right)\left(5.89 \times 10^{-7} \text{ m}\right)}$$

$$= 2.1 \frac{\text{joule}}{\text{coulomb}} = 2.1 \text{ volt.}$$

21. *The ionisation potential of an atom is 14.2 eV. The series limit in its absorption spectrum is calculated. Datas as in last problem.*

Solution : The series limit in the absorption spectrum of an atom corresponds to the energy which when absorbed by a ground state atom, ionises it. Thus, if V be the ionisation potential of an atom, the wavelength at the series limit is given by

$$\lambda = \frac{hc}{eV}$$

$$= \frac{\left(6.62 \times 10^{-34} \text{ joule-sec}\right) \times \left(3 \times 10^{8} \text{ m-sec}^{-1}\right)}{\left(14.2 \times 1.6 \times 10^{-19} \text{ joule}\right)}$$

$$= 0.874 \times 10^{-7} \text{ m}$$

$$= 874 \text{ Å}.$$

22. *The recoil speed of hydrogen atom after it emits a photon in going from n = 3 to n = 1 state is found out in following way. Electron mass is 9.11 × 10^{-31} kg and h — 6.626 × 10^{-34} J-s.*

Solution: The energy of the hydrogen atom in the *nth* state is given by

$$E_n = -\frac{Rhc}{n^2}$$

where R is Rydberg constant. From this, we get

$$E_1 - E_2 = -\frac{Rhc}{1^2} + \frac{Rhc}{3^2} = -\frac{8}{9} Rhc.$$

The energy of the emitted photon is

$$\Delta E = E_1 \sim E_3 = \frac{8}{9} Rhc.$$

The momentum of the photon is

$$p = \frac{\Delta E}{c} = \frac{8}{9} Rh.$$

By conservation of momentum, the recoil momentum of the hydrogen atom will be equal (and opposite) to the momentum of the emitted photon. The recoil speed of the atom is

$$v = \frac{\text{momentum}}{\text{mass}} = \frac{8}{9}\frac{Rh}{m_H}.$$

But $m_H = 1836\ m$, where m is electron mass.

$$\therefore \quad v = \frac{8}{9}\frac{Rh}{(1836\ m)}.$$

Putting the given values of h, m and using $R = 1.097 \times 10^{-7}$ meter$''^{1}$, we get

$$v = \frac{8}{9}\frac{\left(1.097 \times 10^{7}\ \text{m}^{-1}\right)\left(6.626 \times 10^{-34}\ \text{J-s}\right)}{1836\left(9.11 \times 10^{-31}\ \text{kg}\right)}$$

$$= 3.86\ \text{m/s}.$$

23. ***A positronium atom is a system consisting of a positron and an electron. The reduced mass, the Rydberg constant and the wavelength of the first Balmer line for positronium are calculated as follows. (Given m = 9.1 × 10^{-3l} kg, R$_H$= l.09737 × 10^{-3} Å ~ ^{1} and H$_\alpha$ = 1.6563 Å).***

Solution: The positron has the same mass m as the electron and has equal but positive charge. The reduced mass of the electron-positron atom is therefore

$$\mu = \frac{(m)(m)}{m+m} = \frac{1}{2}m = 4.55 \times 10^{-31}\ \text{kg},$$

while the reduced mass of electron in hydrogen is very nearly m.

The Rydberg constant $\left(\frac{\mu e^4}{8 \epsilon_0^2 ch^3}\right)$ for positronium is therefore half than that for hydrogen (with infinitely heavy nucleus). Thus

$$R_p = \frac{1}{2} R_H = 0.54868 \times 10^{-3} \text{ Å}^{-1}.$$

The wavelength of first Balmer line (H_α) for hydrogen atom is given by

$$\frac{1}{\lambda_H} = R_H \left(\frac{1}{2^2} - \frac{1}{3^2}\right),$$

while that for positronium atom is

$$\frac{1}{\lambda_p} = R_p \left(\frac{1}{2^2} - \frac{1}{3^2}\right),$$

Thus $$\frac{\lambda_p}{\lambda_H} = \frac{R_H}{R_p} = 2.$$

$\therefore$ $$\lambda_p = 2\lambda_H = 2 \times 6563 = 13126 \text{ Å}$$

24. *A μ meson (charge — e, mass = 207 m, where m is mass of electron) can be captured by a proton to form a hydrogen-like "mesic" atom. The radius of the first Bohr orbit, the binding energy, and the wavelength of the first time in the Lyman series for such an atom are calculated here. The mass of the proton is 1836 times the mass of electron. The radius of first Bohr orbit and the binding energy of hydrogen are 0.53 Å and 13.6 eV respectively. R$_\infty$ = 109737 cm^{-1}.*

Solution: The reduced mass of the system is

$$\mu = \frac{(207\, m)(1836\, m)}{207\, m + 1836\, m} = 186\, m.$$

From Bohr theory, the radius of the first orbit ($n = 1$) of a hydrogen-like atom for Z–1, is given by (taking finite mass of nucleus in consideration)

$$r_1 = 4\pi\epsilon_0 \frac{h^2}{4\pi^2 \mu e^2} = 4\pi\epsilon_0 \frac{h^2}{4\pi^2 (186\, m)e^2}$$

$$= \frac{1}{186}\left(4\pi\epsilon_0 \frac{h^2}{4\pi^2 me^2}\right)$$

The quantity in the bracket is the first Bohr orbit of hydrogen atom which is 0.53 Å. Therefore

$$r_1 = \frac{1}{186} \times 0.53 \text{ Å} = 2.85 \times 10^{-3} \text{ Å}.$$

Again, from Bohr's theory, the ground-state energy for a hydrogen-like atom with Z = l is given by

$$E_1 = -\frac{\mu e^4}{8\epsilon_0^2 h^2} = -186 \frac{me^4}{8\epsilon_0^2 h^2}$$

$$= -186 \times 13.6 = -2530 \text{ eV}.$$

Hence the binding energy is 2530 eV.

The wavelengths of the Lyman lines are given by

$$\frac{1}{\lambda} = R_M \left(\frac{1}{1^2} - \frac{1}{n^2}\right), \qquad n = 2, 3, 4, \ldots$$

where R_m is she Rydberg constant for the mesic atom. For the first line, $n = 2$ so that

$$\lambda = \frac{4}{3R_M}.$$

Now, $R_M = \dfrac{\mu e^4}{8\epsilon_0^2 ch^3}$ and $R_\infty = \dfrac{me^4}{8\epsilon_0^2 ch^3}$, so that

$$R_M = \frac{\mu}{m} R_\infty = 186\, R_\infty = 186 \times 109737 \text{ cm}^{-1}.$$

Hence $\lambda = \dfrac{4}{3R_M} = \dfrac{4}{3 \times 186 \times 109737\ \text{cm}^{-1}}$

$= 6.52 \times 10^{-3}$ cm = 6.52 Å.

25. ***The wavelength of the first Balmer line (H_α) of hydrogen is 6562.8 Å. The wavelength of the same line for tritium ($_1H^3$) is calculated as follows.***

Solution: The wavelengths of the H_a line ($n = 3 \rightarrow n = 2$) of hydrogen ($_1H^1$) and tritium ($_1H'^3$) are given by

$$\frac{1}{\lambda_H} = R_H \left(\frac{1}{2^2} - \frac{1}{3^2}\right) = \frac{5}{36} R_H$$

and
$$\frac{1}{\lambda_T} = R_T \left(\frac{1}{2^2} - \frac{1}{3^2}\right) = \frac{5}{36} R_T,$$

where R_H and R_I are Rydberg constants for hydrogen and tritium respectively (Z = 1 for both). From this, we get

$$\frac{\lambda_T}{\lambda_H} = \frac{R_H}{R_T}. \quad \text{... } (i)$$

We know that the Rydberg constant for an atom of mass M is given by

$$R_M = \frac{R_\infty}{1 + \dfrac{m}{M}},$$

where R_∞ is the Rydberg constant for *any* atom of infinitely heavy nucleus. Thus, eq. (*i*) gives

$$\frac{\lambda_T}{\lambda_H} = \frac{1 + \dfrac{m}{M_T}}{1 + \dfrac{m}{M_H}},$$

Now $M_T = M_H$ and $M_H = 1836\ m$. Thus

$$\frac{\lambda_T}{\lambda_H} = \frac{1+\dfrac{1}{3\times 1836}}{1+\dfrac{1}{1836}} = \frac{5509}{5511}.$$

But $\lambda_H = 6562.8$ Å.

$$\therefore \quad \lambda_T = \frac{5509}{5511} \times 6562.8 \text{ Å} = 6560.4 \text{ Å}.$$

PROBLEMS IN THE C.G.S. SYSTEM

Note : The formulae deduced in this chapter belong to M.K.S. System. The values of h, e, m, c and R used in the above problems are also in the M.K.S. system. The formulae in the C.G.S. system are as follows : (They can be deduced exactly in the same manner as above by writing the force of electrostatic attraction between the electron and the nucleus as $\frac{e^2}{r^2}$ instead of $\left(\frac{1}{4\pi \in_0}\frac{e^2}{r^2}\right)$:

$$r = \frac{n^2h^2}{4\pi^2 me^2}$$

$$E = -\frac{2\pi^2 me^4}{h^2}\left(\frac{1}{n^2}\right)$$

$$\frac{1}{\lambda} = \frac{2\pi^2 me^4}{ch^3}\left(\frac{1}{n_f^2}-\frac{1}{n_i^2}\right)$$

$$= R\left(\frac{1}{n_f^2}-\frac{1}{n_i^2}\right),$$

where $R = 2\pi^2 me^4/ch^3$ (Rydberg constant).

In these formulae we shall use the following values :

$h = 663 \times 10^{-27}$ erg-sec, $e = 48 \times 10^{-10}$ esu, $m = 9.1 \times 10^{-28}$ gm, $c = 3.0 \times 10^{10}$ cm/sec, $R = 1.097 \times 10^5$ cm^{-1}, 1 eV $= 1.6 \times 10^{-12}$ erg, 1 Å $= 10^{-8}$ cm.

1. ***The hydrogen atom (i) radius of the first orbit, (ii) velocity of the electron in the ground state, (iii) Rydberg constant are calculated. (h = 6.63 × 10⁻²⁷ erg-sec, m = 9.1 × 10⁻²⁸ gm, e = 4.8 × 10⁻¹⁰ esu, c — 3.0 × 10¹⁰ em/sec and proton mass M_p = 1.67 ×10⁻²⁴ gm)***

Solution : Let v cm/sec be the velocity of electron (mass m gm, charge e esu) in an orbit of radius r cm. The positive charge on the nucleus is *Ze,* where Z is atomic number. In case of hydrogen atom, Z = 1, so that positive charge on the nucleus is *e.* As the centripetal force is provided by the electrostatic attraction, we have

$$\frac{nv^2}{r} = \frac{e^2}{r^2}$$

or $$mv^2 = \frac{e^{2,}}{r}. \quad \text{... } (i)$$

From Bohr's postulate, the angular momentum of the electron is given by

$$mvr = \frac{nh}{2\pi}, \quad \text{... } (ii)$$

where n is an integer. Squaring (*ii*) and dividing by (*i*), we get

$$r = \frac{n^2h^2}{4\pi^2me^2}. \quad \text{... } (iii)$$

For the first orbit, n = 1.

$$\therefore \quad r = \frac{h^2}{4\pi^2me^2}$$

$$= \frac{\left(6.63 \times 19^{-27}\right)^2}{4 \times (3.14)^2 \times \left(9.1 \times 10^{-28}\right) \times \left(4.8 \times 10^{-10}\right)^2}$$

$$= 0.53 \times 10^{-8} \text{ cm} = 0.53 \text{ Å}.$$

(*ii*) Dividing eq. (*i*) by (*ii*), we obtain

$$v = \frac{2\pi e^2}{nh}.$$

For the ground state, $n = 1$. Therefore

$$v = \frac{2\pi e^2}{h}$$

$$= \frac{2 \times 3.14 \times \left(4.8 \times 10^{-10}\right)^2}{6.63 \times 10^{-27}}$$

$$= 2.2 \times 10^8 \text{ cm/sec.}$$

(*iii*) The Rydberg constant is $R = \dfrac{2\pi^2 me^4}{ch^3}$. If we take the nuclear motion into account, then the Rydberg constant strictly for hydrogen is given by

$$R = \frac{2\pi^2 me^4}{ch^3}\left(\frac{1}{1 + \left(m / M_p\right)}\right)$$

where M_p is the mass of proton (*i.e.* the hydrogen nucleus). Substituting the values

$$R_H = \frac{2 \times (3.14)^2 \times \left(9.1 \times 10^{-28}\right) \times \left(4.8 \times 10^{-10}\right)^4}{\left(3.0 \times 10^{10}\right) \times \left(6.63 \times 10^{-27}\right)}\left(\frac{1}{1 + \dfrac{9.1 \times 10^{-28}}{1.67 \times 10^{24}}}\right)$$

$$= 1.096 \times 10^5 \text{ cm}^{-1}$$

2. ***The wavelength of the first member and the series limit for the Lyman series of hydrogen is calculated below, ($h = 6.62 \times 10^{-27}$ erg-sec ; $m = 9.1 \times 10^{-28}$ gm, $e = 4.8 \times 10^{-10}$ esu ; $c = 3.0 \times 10^{10}$ cm/sec).***

Solution : According to Bohr's theory of hydrogen atom, the wavelengths of the Lyman series are given by

$$\frac{1}{\lambda} = R\left(\frac{1}{1^2} - \frac{1}{n^2}\right),$$

where R is the Rydberg's constant for hydrogen and $n = 2, 3, 4, 5..$

Now $$R = \frac{2\pi^2 me^4}{ch^3}$$

$$= \frac{2 \times (3.14)^2 \times \left(9.1 \times 10^{-28}\right) \times \left(4.8 \times 10^{-10}\right)^4}{\left(3.0 \times 10^{10}\right) \times \left(6.62 \times 10^{-27}\right)^3}$$

$$= 1.097 \times 10^5 \text{ cm}^{-1}.$$

Hence $$\frac{1}{\lambda} = 1.097 \times 10^5 \left(\frac{1}{1^2} - \frac{1}{n^2}\right) \text{ cm}^{-1}.$$

For the first line, $n = 2$.

$$\therefore \quad \frac{1}{\lambda} = 1.097 \times 10^5 \left(\frac{1}{1^2} - \frac{1}{2^2}\right)$$

$$= 1.097 \times 10^5 \times \frac{3}{4} \text{ cm}^{-1}$$

or $$\lambda = \frac{4}{1.097 \times 10^5 \times 3} = 1.215 \times 10^{-5} \text{ cm}$$

$$= 1215 \times 10^{-8} \text{ cm} = 1215 \text{ Å}.$$

For the series limit, $n = \infty$.

$$\therefore \quad \frac{1}{\lambda_\infty} = 1.097 \times 10^5 \left(\frac{1}{1^2} - \frac{1}{\infty}\right) = 1.097 \times 10^5$$

or λ_∞. $= 0.912 \times 10^{-5}$ cm $= 912 \times 10^{-8}$ cm $\doteq 912$ Å.

3. The change in energy when an electron in the hydrogen atom jumps from the first orbit to second is calculated ($e = 4.8 \times 10^{-10}$ esu, $h = 6.62 \times 10^{-27}$ erg-sec, $m = 9.1 \times 10^{-28}$ gm, 1 eV $= 1.6 \times 10^{-12}$. erg).

Solution : The energy of an electron in nth orbit of hydrogen atom is given by

$$E_n = -\frac{2\pi^2 me^2}{n^2h^2}$$

The change in energy when electron jumps from $n = 1$ to $n = 2$ orbit is

$$\Delta E = E_2 - E_1 = -\frac{2\pi^2 me^4}{4h^2} - \left(-\frac{2\pi^2 me^4}{h^2}\right)$$

$$= \frac{3\pi^2 me^4}{2h^2}$$

$$= \frac{3 \times (3.14)^2 \times \left(9.1 \times 10^{-28}\right) \times \left(4.8 \times 10^{-10}\right)^4}{2 \times \left(6.62 \times 10^{-27}\right)^2}$$

$$= 1.63 \times 10^{-11} \text{ erg}$$

$$= \frac{1.63 \times 10^{-11}}{1.6 \times 10^{-12}} = 10.2 \text{ eV.}$$

9

Cosmic Rays

Elementary Particles

Role of Cosmic Rays

Cosmic rays are highly penetrating radiations consisting of high-energy atomic nuclei which are continuously coming from outer space. About 10^{18} of them reach the earth each second. They have a wide range of energy from 10^9 to 10^{18} electron-volts.

Elster and Geital in 1899 and Wilson in 1900 first detected the cosmic rays. They found that the charge on a very well-insulated electroscope always 'leaked' away in time although there was no ionising agent nearby which could make the air conducting. It was then thought that the leakage of charge was due to the ionising radiation from radioactive minerals in the earth.

Hess, in 1911, sent electroscope up in ballon and found that the intensity of the unknown ionising radiation was larger at high altitudes than at the surface of the earth. He therefore proposed that the radiation originated from somewhere outside the earth's atmosphere. The radiation was finally called 'cosmic rays' by Millikan in 1925.

From 1926 to 1928, Millikan and Cameron conducted experiments on the penetration of cosmic rays through water by lowering the electroscope in deep lakes. They found that the cosmic rays reaching the earth consisted of two components, soft and hard, of very different penetrating powers.

The soft component (now known to consist of electrons, positrons and photons) was absorbed at a depth of about 1 meter of water, but the more penetrating hard component (now known to consist of mesons) was not fully absorbed even at the bottom of the 500-meter deep lake. This showed that the energy of the cosmic rays was higher than any radiation known at that time.

Primary and Secondary Cosmic Rays : Investigations have shown that there are two types of cosmic rays : primary cosmic rays and secondary cosmic rays.

The primary cosmic rays are those which come from outer space and enter the upper part of the earth's atmosphere. They consist of atomic nuclear which have lost their orbital electrons. Most of them are hydrogen nuclei (protons), some are helium nuclei (α-particles), and a few are the nuclei of heavier elements. Their energies are distributed over a wide range.

The secondary cosmic rays are produced in the upper atmosphere by the collision of primary rays with the nuclei of the atmospheric gases. Thus *the rays found near earth's surface are the secondary cosmic rays.* These rays consist chiefly of μ-mesons, and positive and negative electrons. When a primary cosmic-ray proton collides with an air nucleus in the upper atmosphere, it produces mostly π-mesons (positive, negative and neutral) and some hyperons.

The air nucleus is left in an excited state from which the remaining nucleons (protons and neutrons) come out. These secondary nucleons and charged π-mesons multiply further by undergoing similar nuclear collisions.

Many of the charged π-mesons decay into μ-mesons :

$$\pi^{\pm} \rightarrow \mu^{\pm} + v \qquad \text{(neutrino).}$$

The μ-mesons do not interact with nuclei and arrive at sea level with very little energy loss. Many of them even penetrate deep into the ground. Hence *at sea level, about 80% of the (secondary) cosmic rays are μ-mesons.*

Cosmic-ray Showers : The neutral π-mesons produced in the upper atmosphere decay into γ-ray photons :

$$\pi^{o} \rightarrow \gamma + \gamma.$$

These γ-rays in turn produce pairs of positive and negative electrons :

$$\gamma \rightarrow e^{+} + e^{-}.$$

The electrons radiate more photons which, in turn, produce pairs. Thus a cascade shower process continues, whose alternate links are electron pair and photon. Such showers were actually seen by Blackett and Occhialini in 1933 in cloud-chamber photographs.

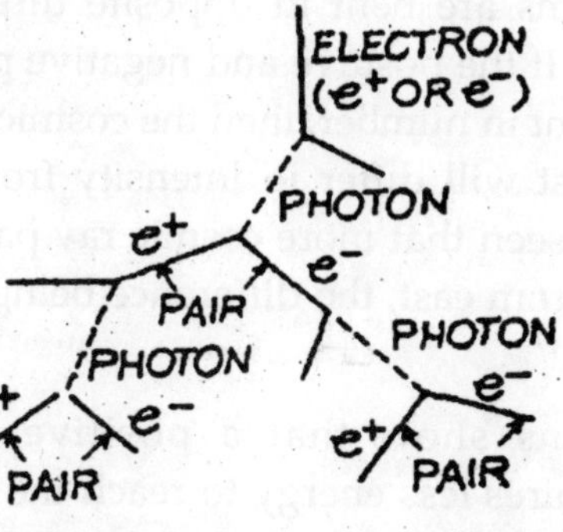

Variations in Cosmic Rays : The intensity of cosmic rays varies with latitude, direction and altitude.

Change of Latitude

The intensity of cosmic rays is greatest and practically uniform in the regions of earth's magnetic poles ; it decreases more and more rapidly as the equator is approached.

This variation is easily explained. As the primary cosmic rays, which are positively-charged particles, approach the earth's atmosphere vertically, their paths are deflected away from the normal by the earth's magnetic field. This deflection varies with magnetic latitude. Particles falling along the magnetic axis of the earth (towards the magnetic poles) do not suffer any deviation and reach the earth even if they have very small energy.

Therefore, the number of cosmic-ray particles reaching the magnetic poles is maximum. Particles falling at an angle with the earths magnetic axis describe a helical path, and those moving very slowly may be bent so much that they do not reach the earth's surface. Those arriving on the magnetic equator suffer the largest deflection because they are moving perpendicular to the magnetic field. So, only most energetic particles can reach the earth at the equator.

Directional Variation : East-West Asymmetry : Particles of opposite signs are bent in opposite directions by earth's magnetic field. If the positive and negative particles in cosmic rays are different in number, then the cosmic rays reaching the earth from west will differ in intensity from those reaching from east. It is seen that more cosmic ray particles come from the west than from east, the difference being maximum at the equator.

Calculations show that a positive particle coming from west requires less energy to reach the equator than that coming from the east. Since primary cosmic rays are positively-charged, they are expected more frequently from west than from east.

Change of Altitude

The intensity of cosmic rays increases with altitude. It, however, reaches a maximum not at the top of the atmosphere but well below it, above which the intensity again drops.

The position of the maximum indicates that much of the cosmic radiation must actually be produced there. This happens as the primary rays suffer nuclear collisions, producing secondary rays.

Birth of Cosmic Rays

The origin of primary cosmic rays is still an unsolved problem. We do not know definitely where cosmic rays come from and how they acquire their tremendous energies.

One of the views is that some of the cosmic rays may come from the sun. According to this view, some of the protons in the sun acquire high energies at times of solar activity when great eruptions take place and masses of ionised gases shoot out from the sun into interplanetary space. The main argument against this assumption is the uniform intensity of cosmic rays at all hours of the day and night. Counter-arguments are, however, also in progress.

Another view is that the cosmic rays come from the stars. This view is based on the fact that the total energy flux at the top of the atmosphere due to cosmic rays is about equal to the radiant energy flux received from all stars excluding the sun. This view is also unacceptable because every star cannot emit equally the light and the cosmic rays.

It has also been suggested that some special class of stars known as 'flare stars', 'magnetic variables' and 'supernova' may be responsible for most of the cosmic rays.

Most scientists believe, however, that cosmic-ray particles are accelerated by electro-magnetic processes as take place in solar flares. Stars have been observed in which eruptions resembling solar flares occur. Double stars with large magnetic moments and revolving around a common centre have also been observed. All of these are possible sources of cosmic rays.

Elementary Particles : *An elementary particle is one whose internal structure cannot be described as a combination of other particles.*

Until 1932, the only elementary particles known were the positively-charged proton (p), the negatively-charged electron (e) and the photon (γ). In that year two new elementary particles were discovered, namely, the neutron (n) and the positive electron or positron (e^+).

From 1932 onwards the number of elementary particles known has increased continuously. The negatively and positively charged mu-mesons or muons (μ^- and μ^+) were discovered in 1936, charged pi-mesons or pions (π^+ and π^-) in 1947, a neutral pion (π°) in 1950 and the negatively-charged counterpart of the proton known as antiproton $(\overline{p})$ in 1955. Then came the antineutron $(\overline{n})$, a new family of charged and neutral mesons known as *K*-mesons or kaons (K^+, K^-, K°), and hyperons (particles heavier than proton). The hyperon family includes lambda-hyperon (δ°), sigma-hyperons (Σ^+, Σ^-, Σ°), xi-hyperons (Ξ^-, Ξ°) and omega-hyperon (Ω°).

The existence of a neutral, massless particle neutrino (v) was suggested by Pauli in 1930 and theoretically established by Fermi in 1934. The particle was actually discovered in 1956. Further in 1963, it became known that there are two kinds of neutrinos, the electron-neutrino (v_e) and the muon-neutrino ($v\mu$).

Characteristic Properties of Elementary Particles : The elementary particles are characterised by their mass, charge, life-time, spin, etc.

Mass : An elementary particle has always the same rest mass. If two particles have different rest masses, we consider them different particles. The magnitude of the rest mass serves as the principal label which identifies the particle uniquely. The value of the rest mass may allow us to infer the existence of a particle even without a direct observation, just from the conservation of energy and momentum. The neutral pion π° was discovered in this very way.

Charge : All known elementary particles have charge $+e, -e$, or zero. Further, the charge is always conserved in any collision process. This is clear in the following neutron-proton collisions :

$$n + p \rightarrow p + p + \pi^{-}.$$

$$n + p \rightarrow n + n + \pi^{+}$$

$$n + p \rightarrow n + p + K^{-} + K^{+}.$$

Average Lifetime : All elementary particles, except photon, electron, proton and neutrinos are unstable and undergo decay into other elementary particles of smaller mass. The decay probability of a particular particle is, however, independent of the length of line the particle has lived. It is impossible to predict when a given (unstable) particle will decay. Hence an elementary (unstable) particle has an *average* life-time, which is independent of the way the particle decays.

Spin : Many elementary particles spin in a manner analogous to that of the earth on its axis, but with certain differences. The spin property forms a basis for the classification of elementary particles.

Interactions : Four kinds of interactions between elementary particles are known : *gravitational, weak, electromagnetic and strong*. Their relative magnitudes are in the ratios 10^{-39}; 10^{-13}; 10^{-3}: 1. The (weakest) gravitational interaction is universal. All elementary particles without any exception are subjected to it. It is the interaction that holds the earth together, binds the sun and planets into the solar system, and binds stars into galaxies. It is, however, insignificant in dealing with atoms and nuclei.

The weak interaction causes the light particles to interact with one another and with heavier particles. It is responsible for particle decays in which neutrinos are involved, notably beta decays. The electromagnetic interaction binds electrons to atoms and binds atoms together to form molecules and crystals.

The strong interaction is found between heavier particles. For example, it binds neutrons and protons to form the nuclei of all the elements. It is the dominant interaction in high-energy particle physics.

Antiparticles : The Dirac's relativistic quantum theory of electron predicted the existence of what is called an antiparticle. The antiparticle of a given particle has exactly the same mass, spin and life-time (if unstable) but *opposite* charge (if any). Further, the alignment between the spin and magnetic moment of an antiparticle is *opposite* to that of the corresponding particle. Further, when an antiparticle meets its particle, they mutually annihilate.

The first antiparticle known was the positron which was discovered in 1932 in a cloud chamber exposed to cosmic rays. It is a particle having the same mass as an electron, but opposite charge. When a positron comes to rest in matter, it is quickly annihilated by an electron resulting in two γ-photons :

$$e^+ + e^- \rightarrow 2\gamma.$$

The γ-photons fly away in different directions such as to conserve both momentum and energy.

Conversely, when a γ-photon of energy greater than 1.02 MeV passes near a nucleus in a substance, an electron and a positron are created (pair production):

$$\gamma + X \rightarrow e^- + e^+ + X,$$

where X is the nucleus in whose force field the pair is born. Any available energy in excess of 1.02 MeV goes into the kinetic energies of the electron and the positron. The presence of a nucleus X is required in order that momentum as well as energy be conserved in the process. The nucleus receives the excess momentum of the γ-photon over the total momentum of an electron and a positron.

The antiparticle of a proton is called the antiproton (or negative proton) and is denoted by $\bar{p}$. It was first produced in 1955 in proton-proton and proton-neutron collisions :

$$p + p \rightarrow p + p + p + \bar{p}$$

$$p + n \rightarrow p + n + p + \bar{p}$$

The energy of the bombarding proton on metal targets was of the order of 6 GeV (1 GeV = 10^9 eV). An antiproton differs from a proton in the sign of its electric charge and in its magnetic moment (in an antiproton the magnetic moment is directed oppositely to the mechanical angular momentum).

Antineutrons ($\bar{n}$) were observed in 1956 in certain antiproton-proton collisions :

$$p + \bar{p} \rightarrow n + \bar{n} .$$

An antineutron differs from a neutron in the sign of its instrinsic magnetic moment (in an antineutron the direction of the magnetic moment coincides with that of the mechanical angular momentum).

The annihilation of a particle-antiparticle pair need not always result in a pair of γ-photons (energy), as it does in the case of electron-positron annihilation. When an antiproton is annihilated with a proton, for example, several neutral and charged re-mesons are usually produced :

$$p + \bar{p} \rightarrow \pi^+ + \pi^- + \pi^+ + \pi^- + \pi^o$$

$$p + \bar{p} \rightarrow \pi^+ + \pi^- + \pi^o + \pi^o + \pi^o$$

An antiproton may annihilate when it encounters not only a proton, but also a neutron. For example, the following process is possible :

$$\bar{p} + n \rightarrow \pi^+ + \pi^- + \pi^- + \pi^o + \pi^o$$

Similar annihilations are observed for neutrons.

The distinction between the neutrino and the antineutrino is particularly interesting. The neutrino spins counter clockwise when viewed from behind, while antineutrino spins clockwise.

There are particles that are identical with their antiparticle (*i.e.* they have no antiparticles). Such particles are called 'absolutely neutral'. These include the photon, π^o-meson and the η-meson. These particles are not capable of annihilation. All the other particles do have their antiparticles.

Classification of Elementary Particles : Elementary particles are usually classified into four groups according to their mass and spin. These are :

1. *Photon (γ) :* Photon forms a group by itself. It is a stable particle with zero charge and zero rest mass. It has a spin equal to 1, and is therefore a boson. It is the quantum of the electromagnetic field and participate in electromagnetic interactions.
2. *Leptons :* These are light-weight particles. They include electrons (e^-, e^+), muons (μ^-, μ^+), electron-neutrinos (v_e, $\bar{v}_e$) and muon-neutrinos (v_μ, $\bar{v}_\mu$). All leptons have a spin equal to ½, and are therefore fermions. They interact weakly with other particles. Except muons, the leptons are stable.
3. *Mesons :* These are middle-weight particles having masses intermediate between the muons and the nucleons (protons and

Class	Name	Particle symol	Anti-particle	Spin	Restmass (m_e)	average-life (sec)
Photon	Photon	γ	(γ)	1	0	stable
Leptons	Eloectron	e	e	1/2	1	stable
	μ-meson	μ^-	μ	1/2	207	2.2×10[illegible]
	e-neutrino	v_e		1/2	0	stable
	μ-neutrino	v_μ		1/2	0	stable
Mesons (HADRONS)	Positive π-meson	π^+	π	0	273	2.6×10[illegible]
	Neutral π-meson	π^o	(π^o)	0	264	9×10[illegible]
	Positive K-meson	K	K^-	0	966	1.2×10[illegible]
	Neutral K-meson	K^o		0	974	8.6×10[illegible] & 5.2×10[illegible]
	η–meson	η^o	(η^o)	0	1073	2.4×10[illegible]
Baryons (HADRONS)	Nucleons { Proton	p		1/2	1836	stable
	Nucleons { Neutron	n		1/2	1839	930
	Lambda hyperon	$\wedge^o$		1/2	2182	2.5×10^{-10}
	Positive sigma--hyperon	Σ^-		1/2	2328	8.0×10[illegible]
	Neutral " "			1/2	2332	< 10[illegible]
	Negative " "	Σ		1/2	2341	1.5×10[illegible]
	Negative xi hyperon	Ξ		1/2	2571	1.7×10[illegible]
	Neutral " "	Ξ^o		1/2	2583	3.0×10^{-10}
	Negative omega--hyperon	Ω^-		3/2	3290	1.3×10[illegible]

neutrons). They include pions (π^+, π^-, π^o), kaons (K^+, K^-, K^o, $\overline{K^o}$), and the eta-meson (η^o). The spin of all mesons is zero, so that they are bosons. All mesons are unstable.

4. *Baryons* : These are the heavy-weight particles which include nucleons (p, $\overline{p}$, n, $\overline{n}$) and hyperons ($\wedge^0$, Σ^+, Σ, Σ^o, Ξ^-, Ξ^o, Ω^-). The spin of all baryons is ½ except that of the Ω hyperon, which is $\frac{3}{2}$. Thus baryons are fermions. Except for the proton, all baryons are unstable. When a baryon decays, another baryon is formed without fail in addition to other particles. This is one of the manifestations of the law of baryon charge conservation.

Mesons and baryons are strongly-interacting particles, and collectively they are called 'hadrons'.

The classifying table shows particles as well as antiparticles. When the antiparticle is designated with the aid of a bar (—), the name of the antiparticle is obtained by adding the prefix "anti" to the name of the relevant particle. For example, the antiparticle of the lambda hyperon is called antilambda hyperon. The antiparticle of the electron is the positron. In the remaining cases, the names of particles and antiparticles are distinguished by adding the words 'positive' and 'negative'. For example, the negative pi-meson is the antiparticle of the positive pi-meson. Those antiparticles which are identical with their particles are shown in brackets.

Besides above, there are elementary particles predicted theoretically but not yet discovered, e.g., quarks which have fractional charges.

Leptons : These are light-weight elementary particles which are stable, or decay only by weak interactions. They are all fermions (having an intrinsic spin of ½) and occur as particles and antiparticles. Leptons include electrons, μ-mesons, electron-neutrinos and muon-neutrinos and their antiparticles.

(i) *Electrons and Positrons :* Electron is a stable atomic particle of mass of 9.1×10^{-31} kg and negative charge of 1.6×10^{-19} coulomb. Its antiparticle positron is identical with electron except that it is positively charged. When electron and positron meet, they annihilate each other, producing two γ-photons :

$$e^{+} + e^{-} \rightarrow \gamma + \gamma.$$

(ii) *Ma-mesons (or Muons) :* Mu-mesons, discovered in 1936 by Anderson and Neddermeyer, are the components of cosmic radiation which are found in abundance in the cosmic rays at the ground level. They are created as a result of pi-meson decay in cosmic radiation.

Mu-mesons exist as both negative and positive and are denoted as μ^{-} and μ^{+} respectively. Both have the same rest mass of 207 m_c. (intermediate between the rest mass of an electron and the rest mass of a proton) and the same spin of ½. The μ^+ meson is the antiparticle of μ^- meson. There is no neutral mu-meson.

Both μ^- and μ^+ mesons are unstable, having an average life of 2.2×10^{-6} sec. They decay spontaneously into electron (or positron) according to the following scheme :

$$\mu^- \rightarrow e^- + \overline{v}_e + v_\mu$$

$$\mu^+ \rightarrow e^+ + v_e + \overline{v}_\mu .$$

The neutrino and antineutrino are not actually observed, their emission is postulated to conserve both energy and momentum. An energy of 105 MeV is released in this decay.

Unlike pi-mesons (which interact strongly with nuclei), the μ-mesons interact *weakly* with nuclei. Hence they penetrate considerably through matter before being absorbed. Generated high in the atmosphere, most of them reach sea-level and some penetrate considerable distances into the earth. Hence the hard component of secondary cosmic rays consist mostly of μ-mesons.

The negative mu-meson can replace electron in atomic Bohr orbit to form 'mesic' (or 'muonic') atom. Thus, apart from its extra mass, a mu-meson behaves like an electron.

(iii) *Electron-neutrinos and Muon-neutrinos :* These particles, denoted by v_e and v_μ, have negligible rest mass and no charge, and are spin ½ fermions. They are distinguished from their anti-particles $\overline{v}_e$, and $\overline{v}_\mu$ by their *helicities* which are left-handed for neutrinos and right-handed for antineutrinos. They participate in weak interaction with matter.

Mesons : These are middle-weight particles having masses intermediate between the muons and the nucleons (protons and neutrons). They are all bosons having zero spin and are unstable. They include pions (π^+, π^-, π°), kaons ($K^+, K^-, K^\circ, \bar{K}^\circ$) and the eta-meson ($\eta^\circ$).

(i) *Pi-mesons (or Pions)* : Pi-mesons were discovered in 1947 in the upper atmosphere, although their existence was predicted much earlier by Yukawa in his meson theory of nuclear forces. They are created in nature when primary cosmic-ray protons collide with atmospheric nuclei and cause energetic nuclear disintegrations. After the development of high-energy accelerators, pi-mesons have also been created artificially.

Pi-mesons can exist in all three states; π^+, α^- and π°, all having a zero spin. The π^+ and π^- mesons have rest masses of 273 m_e, while that of the π° mesons is slightly less, 264 m_e. The π^- meson is the antiparticle of the π^+ meson, while the π° meson is its own antiparticle.

Pions interact strongly with nucleons. For example, they are produced by collisions of high-energy (several hundred MeV) protons produced in synchrocyclotron. Some pion production reactions are

$$p + p \rightarrow p + n + \pi^+$$

$$p + n \rightarrow p + p + \pi^-$$

$$p + \gamma \rightarrow n + \pi^+$$

$$p + \gamma \rightarrow p + \pi^\circ$$

The pi-mesons are highly unstable. The average life-time of (charged) π^+ and π^- mesons is only 2.6×10^{-8} sec, while that of (neutral) π° meson is still shorter, 9×10^{-17} sec. These life-times are so short that only a fraction of cosmic ray pi-mesons can reach earth. Charged pi-mesons decay in flight by weak interaction into lighter mu-mesons and mu-neutrinos :

$$\pi^+ \rightarrow \mu + v_\mu$$

$$\pi^- \rightarrow \mu- + \overline{v}_\mu$$

These mu-neutrinos are somewhat different from the neutrinos involved in beta-decay which are better called electron-neutrinos. The emission of the mu-neutrinos is postulated to conserve both energy and momentum.

As the rest mass of a mu-meson is 207 m_e and that of neutrino is zero, a mass-energy of $(273 - 207)m_e = 66\ m_e$ is available in the above decay. As m_e is equivalent to 0.51 MeV, the energy released is $66 \times 0.51 = 33.7$ MeV. Of this energy, a constant amount of 4 MeV appears as the kinetic energy of the mu-meson. The mu-mesons are also unstable and decay into electrons.

The neutral pi-meson decays spontaneously by an electro-magnetic interaction into two high-energy γ photons :

$$\pi^\circ \rightarrow \gamma + \gamma.$$

The energy released is shared equally between the two γ-photons.

(ii) *K-mesons (or Kaons)* **:** There is also a heavier class of mesons, called K-mesons, which were first discovered in high-altitude cosmic ray experiments but now are available from the accelerators. They exist as K^+ and its antiparticle K^-, and also as K° and its antiparticle $\overline{K}^\circ$. The charged K-mesons (K^+ and K^-) have rest masses of 966 m_e, spin of zero, and mean-lives of 1.2×10^{-8} sec.

K-mesons are easily produced through strong interaction. An example is the reaction

$$\pi^- + p \rightarrow \Lambda^\circ + K^\circ$$

The charged meson decay in a variety of ways :

$$K^+ \rightarrow \pi^+ + \pi^+ + \pi^-$$

$$K^\pm \rightarrow \pi^\pm + \pi^\circ$$

$$K^\pm \rightarrow \mu^\pm + \upsilon_\mu$$

There are two distinct varieties of neutral K-mesons, $K_1°$ and $K_2°$. Both have rest masses 974 m_e, but half-lives of the order of 10^{-10} and 10^{-8} sec respectively. They can decay in several modes :

$$K_1° \rightarrow \pi^+ + \pi^-$$

$$\rightarrow \pi^o + \pi^o$$

$$K_2° \rightarrow \pi^+ + \pi^- + \pi^o$$

$$\rightarrow \pi^o + \pi^o + \pi^o.$$

(iii) *η-mesons :* The neutral eta-meson η° was discovered in 1961. It has a rest mass of 1073 m_e, which is intermediate between that of a K-meson and that of a nucleon. Like others, the η-meson is also a zero spin boson. It decays electromagnetically in about 10^{-19} sec into two photons.

Role of Moons and Pions

Muons and pions are elementary particles having rest masses intermediate between the rest mass of an electron and that of a nucleon (proton or neutron). The muon was discovered in 1936 in the cosmic rays at ground level, while the pion was discovered in 1947 also in cosmic rays but in the upper atmosphere although its existence had been predicted by Yukawa in 1935 in connection with nuclear forces.

Although the rest masses of muon and pion are close to each other, but they so much differ in their spin and interaction properties that they have been placed in different classes of elementary particles. The muon is a lepton, while the pion is a meson. Their comparative properties are as under:

	MUON	PION
1. Kinds:	μ^-, μ^+	π^+, π^-, π^o
2. Rest Mass :	207 m_e	273 m_e (π^+, π^-); 264 m_e (π^o)
3. Spin:	½ (fermion)	0 (boson)
4. Interaction with nuclei:	weak	strong

5. Half-life (sec): 2.2×10^{-6} 2.6×10^{-8} (π^+, π^-); 9×10^{-17} (π^o)
6. Decay mode : $\mu^- \rightarrow e^- + \overline{v}_e + v_\mu$ $\pi^+ \rightarrow \mu^+ + v_\mu$

   $\mu^+ \rightarrow e^+ + v_e + v_\mu$ $\pi^- \rightarrow \mu^- + \overline{v}_\mu$

   $\pi^o \rightarrow \gamma + \gamma$

Baryons : These are heavy-weight elementary particles which include nucleons (protons and neutrons) and particles heavier than nucleons. They are all fermions and are strongly interacting. Except proton, all baryons are unstable. The two sub-classes of baryons are :

(i) *Nucleons* : These are the nuclear particles proton (p), neutron (n) and their antiparticles, antiproton $(\overline{p})$ and antineutron $(\overline{n})$. The mass of proton is 1836 m_e and that of neutron is 1839 m_e. They all have a spin of ½.

(ii) *Hyperons* : Elementary particles heavier than nucleons are called 'hyperons'. They are unstable and have an average life-time of the order of 10^{-10} sec. The known hyperons fall into four classes :

  (1) lambda hyperons $\wedge^o$ (with zero charge) and their antiparticles.

  (2) sigma hyperons Σ^+, Σ^o, Σ^-, (with positive, zero and negative charges) and their antiparticles.

  (3) xi hyperons Ξ^-, Ξ^o (with negative and zero charges) and their antiparticles.

  (4) Omega hyperons Ω^- (with negative charge) and its antiparticle.

The spin of all hyperons is ½ except that of Ω-hyperon, which is $\frac{3}{2}$. Some possible decay modes for the hyperons are as follows :

$$\Lambda^0 \rightarrow p + \pi^-$$
$$\rightarrow n + \pi^0$$
$$\Sigma^+ \rightarrow p + \pi^0$$
$$\rightarrow n + \pi^+$$
$$\Sigma^0 \rightarrow \Lambda^0 + \gamma$$
$$\Sigma^- \rightarrow n + \pi^-$$
$$\Xi^- \rightarrow \Lambda^0 + \pi^-$$
$$\Xi^0 \rightarrow \Lambda^0 + \pi^0$$
$$\Omega^- \rightarrow \Xi^0 + \pi^-$$
$$\rightarrow \Xi^- + \pi^0$$
$$\rightarrow \Lambda^0 + K^-$$

Conservation Laws Governing Elementary Particles : The modes of decay of the elementary particles are governed by the conservation laws of *charge, mass-energy, linear momentum and angular momentum* (spin). The application of these laws have led to the prediction of new particles *e.g.* neutrinos.

There are, however, certain modes of decay that are allowed by the above conservation laws, but they are never observed. For example, let us consider the possible modes of proton-decay :

$$p^+ \rightarrow e^+ + \pi^0$$
$$p^+ \rightarrow \mu^+ + \pi^0$$

These modes of decay, however, never occur in nature (proton is a stable particle). Therefore, it is natural to suppose that these are prevented by some unknown laws of conservation. Thus new laws of conservation have been deduced from a study of all possible types of particle decay. These are conservation laws of baryon-number *(B)*, electron lepton-number (*L*) and muon lepton-number *(M)*. These laws plus the old conservation laws help us to determine whether any given decay is capable of taking place or not.

To establish the new conservation laws, three new quantum numbers *B*, *L* and *M* have been introduced.

Baryon Number : Conservation

Each particle is assigned a baryon number *B*, such that

$B = +1$ for all baryons $(p, n, \Lambda^\circ, \Sigma^+, \Sigma^\circ, \Sigma^-, \Xi^-, \Xi^\circ, \Omega^-)$

$B = -1$ for all antibaryons $\left(\overline{p}, \overline{n}, \overline{\Lambda^\circ}, \overline{\Sigma}^+, \overline{\Sigma}^\circ, \overline{\Sigma}^-, \overline{\Xi}^-, \overline{\Xi}^\circ, \overline{\Omega}^-\right)$

$B = 0$ for all others (photon, leptons, mesons). According to the law of conservation, the total value of *B* remains constant in any process involving elementary particles.

Conservation of Electron Lepton-number : There is an electron lepton number *L* for all particles, such that

$L = +1$ for electron (e^-) and electron-neutrino (v_e),

$L = -1$ for antielectron (e^+) and electron-antineutrino $(\overline{v}_e)$

$L = 0$ for all others.

Again, the total value of *L* remains constant in any decay or reaction of elementary particles.

Conservation of Muon Lepton-number : There is a muon lepton-number *M* for all particles, such that

$M = +1$ for muon (μ^-) and muon-neutrino (v_μ)

$M = -1$ for antimuon (μ^+) and muonantineutrino $(\overline{v}_\mu)$

$M = 0$ for all others.

The total value of *M* remains conserved in any process of elementary particles.

As example of the above laws, let us consider the decay of a free neutron:

$$n \rightarrow p + e^- + \bar{v}_e$$

For this B = 1 → 1 + 0 + 0

and L = 0 → 0 + 1 + (–1).

The total value of *B* before and after the decay is 1, and that of *L* is 0. Thus both *B* and *L* are conserved and hence the neutron decay does take place in nature. We know that a free neutron is unstable.

Stability of Proton : The laws of baryon and lepton number conservations are responsible of the stability of the proton. Let us consider a possible mode of a free proton decay :

$$p \rightarrow e^+ + \pi^\circ$$

For this $B = 1 \rightarrow$ 0 + 0

and $L = 0 \rightarrow$ (–1) + 0

The total values of both *B* and *L* are *not* the same before and after the decay, and so the decay is forbidden by both *B* and *L* conservations. Proton decay is really forbidden because there are no baryons of mass smaller than that of proton.

It is for the conservation of the lepton number that in beta-decays an electron is produced together with an antineutrino, and a positron is produced together with a neutrino.

Maximum available Energy for $\mu^- \rightarrow e^- + 2v$: The μ^- meson decays from rest. Therefore, the conservation of linear momentum requires that two particles must be ejected. The only particle observed is the electron. Since the charge is conserved, the second particle must be a neutrino. But since the spin of μ-meson is ½, and the sum of the spins of the electron and neutrino can only be 0 or 1 (but never ½), the conservation of spin (angular momentum) requires another particle of opposite spin ½ to be ejected at the same time. This must be an antineutrino, so that the equation would be

$$\mu^- \rightarrow \quad e^- + v + \bar{v}$$

$$\text{spins} \qquad \frac{1}{2} \rightarrow \quad \frac{1}{2} + 0$$

This raises another difficulty. If the neutrino and antineutrino are ejected simultaneously, they would annihilate in flight to produce two γ-photons ($v + \bar{v}\ 2\gamma$). This, however, is not found.

This shows that v and $\bar{v}$ do *not* form particle-antiparticle pair. Hence v and $\bar{v}$ must be the neutrinos of different species. Two types of neutrinos are therefore postulated : *e*-neutrinos and μ-neutrinos, and the decay equation should now be written as

$$\mu^- \rightarrow e^- + v_\mu + \bar{v}_e \,.$$

Now, taking $M_\mu = 207\ m_e$, the total available energy Q of the reaction is given by

$$207\ m_\theta \rightarrow \ m_e + 0 + 0 \text{ -r } Q,$$

because the mass of neutrinos is zero.

$$\therefore \qquad Q = 206\ m_e.$$

But m_e is equivalent to 0.51 MeV, therefore

$$Q = 206 \times 0.51 = 105 \text{ MeV}.$$

If this energy is equally divided among the three particles, the average energy of the electron is 35 MeV.

PROBLEMS

1. An electron and its antiparticle are annihilated after collision. How much energy will be obtained ? (Mass of electron =9.1 × 10^{-31} kg, speed of light = 3.0 × 10^{8} m/s and Planck's constant = 6.6 × 10^{-34} J-s).

Solution: The positron is antiparticle of electron. In electron-positron collision, both are annihilated and the total mass is converted into two γ-photons (energy) :

$$e^- + e^+ \rightarrow 2\gamma.$$

The mass-loss in this process is

$$\Delta m = 2 \times (9.1 \times 10^{-31}) = 18.2 \times 10^{-31} \text{ kg}.$$

According to mass-energy relation, the energy released is

$$\Delta E = (\Delta m)c^2$$

$$= (18.2 \times 10^{-31} \text{ kg}) \times (3.0 \times 10^8 \text{ m/s})^2$$

$$= 1.64 \times 10^{-13} \text{ joule}$$

$$= \frac{1.64 \times 10^{-13}}{1.6 \times 10^{-19}} \; eV$$

[because 1 eV = 1.6 × 10^{-19} joule]

$$= 1.02 \times 10^6 \text{ eV} = 1.02 \text{ MeV}.$$

This is total energy obtained.

Two γ-photons are produced in this process, hence energy of each photon

$$E = \frac{\Delta E}{2} = \frac{1.64 \times 10^{-13} \text{ joule}}{2}$$

$$= 0.82 \times 10^{-13} \text{ joule} = 0.51 \text{ MeV}.$$

2. A positron and an electron with negligible kinetic energy annihilate each other to produce two photons. What are their frequencies.

Solution: According to the above example, the energy of each of the two photons produced by the union of an electron and a positron is

$$E = 0.82 \times 10^{-13} \text{ joule.}$$

If the frequency of the photon be v, then $E = hv$.

$$\therefore \quad v = \frac{E}{h} = \frac{0.82 \times 10^{-13} \text{ joule}}{6.6 \times 10^{-34} \text{ joule} - \text{second}}$$

$$= 1.2 \times 10^{20} \text{ hertz.}$$

3. ***If a pion decays from rest to give a muon of 4.0 MeV energy, what is the kinetic energy of the accompanying neutrino? What is the mass of the neutrino in this process ?***

Solution: The pion decay is

$$\pi^+ \rightarrow \mu^+ + v_\mu + Q .$$

From this, we have

$$Q = M_\pi - M_\mu \quad \text{(neutrino has zero 'rest' mass)}$$

$$= 273\ m_e - 207\ m_e$$

$$= 66\ m_e$$

$$= 66 \times 0.51 \qquad (\because \ m_e \equiv 0.51 \text{ MeV})$$

$$= 33.7 \text{ MeV.}$$

Since the muon takes 4.0 MeV of energy, the energy shared by the neutrino is 33.7 – 4.0 = 29.7 MeV.

Mass of this neutrino

$$= \frac{29.7}{0.51}\ m_e = 58.2\ m_e.$$

4. ***Find the maximum kinetic energy of the electron emitted in the beta decay of the free neutron. The neutron-proton mass difference is 1.30 MeV.***

Solution: The decay equation is

$$n \rightarrow p^+ + e^- + \overline{v}_e + Q.$$

From this, we have

$$\begin{aligned} Q &= M_n - M_p - m_e \\ &= (1.30 \text{ MeV}) - m_e \\ &= (1.30 \text{ MeV}) - (0.51 \text{ MeV}) \\ &= 0.79 \text{ MeV}. \end{aligned}$$

This energy-release goes into the kinetic energy of the electron plus the kinetic energy of the antineutrino. The two particles can share the 0.79 MeV of energy in any way they choose. Thus the electron may have a maximum kinetic energy of 0.79 MeV.

10

Wave Particles

Wave-Particle and Duality

(de-Broglie Wavelength) : Each material particle in motion behaves as waves and the wavelength λ associated with any moving particle of momentum p is given by

$$\lambda = \frac{h}{p} = \frac{h}{mv}$$

- de-Broglie wavelength of particle of K.E. = E_k is given by

$$\lambda = \frac{h}{\sqrt{2mE_k}}$$

- de-Broglie wavelength for charged particle of charge q and accelerated through a potential difference of V volts is

$$\lambda = \frac{h}{\sqrt{2mqV}}$$

- de-Broglie wavelength of a material particle at temperature T is

$$\lambda = \frac{h}{\sqrt{2mkT}} \qquad k \rightarrow \text{Boltzmann}$$

constant

- de-Broglie wavelength of helium atom

$$\lambda_{He} = \frac{h}{\sqrt{3mkT}}$$

- The relativistic expression for the de-Broglie wavelength of an electron accelerated through a high p.d. of V volts

$$\lambda = \frac{h}{\sqrt{2m_0eV}}\left(1+\frac{eV}{2m_0c^2}\right)$$

Heisenberg's Uncertainty Principle

- The product of uncertainties in determining the position and momentum of a particle at the same instant is

$$\Delta x \, \Delta p \geq \hbar$$

- If ΔE is the uncertainty in determining energy of the system and $\Delta t \rightarrow$ uncertainty in time

$$\Delta E \, \Delta t \geq \hbar$$

- Also $\Delta J \, \Delta\theta \geq \hbar$

  $\Delta J \rightarrow$ Uncertainty in angular momentum

  $\Delta\theta \rightarrow$ Uncertainty in angle

Exact Statement : The product of the uncertainties in determining the position and momentum of the particle can never be smaller than the number of the order $\frac{\hbar}{2}$

$$\therefore \quad \Delta x \, \Delta p \geq \frac{\hbar}{2}$$

Similarly, $$\Delta E \, \Delta t \geq \frac{\hbar}{2}$$

$$\Delta J \, \Delta\theta \geq \frac{\hbar}{2}$$

Applications:

(i) Non-existence of the electrons in the nucleus.

(ii) Radius of Bohr's first orbit.

(iii) Minimum energy of a harmonic oscillator.

(iv) Energy of a particle in one dimensional box.

(v) Finite width of spectral lines.

Schrodinger Wave Equation

(i) Time independent Schrodinger equation

$$\nabla^2\psi + \frac{2m}{\hbar^2}(E-V)\psi = 0$$

(ii) Time dependent Schrodinger equation

$$\left(-\frac{\hbar^2}{2m}\nabla^2 + V\right)\psi = i\hbar\frac{\partial\psi}{\partial t}$$

The operator $\left(-\frac{\hbar^2}{2m}\nabla^2 + V\right)$ is called Hamiltonian represented

by H $\Rightarrow$ while operator $i\hbar\frac{\partial}{\partial t}$, operated on ψ, gives E

$\therefore$ TDSE $\Rightarrow$ Hψ = Eψ

The Commutator : The commutation relation can be written as

$$[xp_x - p_x x] = [x, p_x] = i\hbar$$

The quantity

$$[\alpha, \beta] = \alpha\beta - \beta\alpha = -[\beta, \alpha]$$

is said to be the commutator of the two operators α and β.

Probability Current Density

$$\bar{J} = \frac{i\hbar^2}{2m}[\psi\nabla\psi^* - \psi^*\nabla\psi]$$

and $\psi^*\,\psi \rightarrow$ Position probability density.

- *Expectation values*

$$< x > = \frac{\iiint \psi^* x \, \psi d\tau}{\iiint \psi^* \psi d\tau}$$

Since $\iiint \psi^* \psi d\tau \;=\; 1$

$$\therefore \qquad < x > = \int \psi^* x \, \psi d\tau$$

In general, for any function *f(r)*, the expectation value

$$< f(r) > = \int \psi^* f(x) \, \psi d\tau$$

Since operator, $E \rightarrow i\hbar \frac{\partial}{\partial t}$

$$\therefore \qquad < E > = \int \psi^* i\hbar \frac{\partial}{\partial t} \psi d\tau$$

and operator, $p \rightarrow i\hbar \frac{\partial}{\partial t}$.

$$< p_x > = \int \psi^* \left(-i\hbar \frac{\partial \psi}{\partial x} \right) d\tau$$

$$< p^2{}_x > = \int \psi^* \left(-\hbar^2 \frac{\partial^2 \psi}{\partial x^2} \right) d\tau$$

- ***Ehrenfest's theorem (Relationship between classical and quantum motion)***

$$\frac{d}{dt} < x > = \frac{1}{m} < p_x >$$

$$\frac{d}{dt} < p_x > = -< \frac{\partial V}{\partial x} >$$

- ***One dimensional wave packet : Concepts of group and phase velocities***

A wave packet obtained by the superposition of plane wave

$$\psi(x,t)=\frac{1}{\sqrt{2\pi}}\int_{\infty}^{\infty} A(k)e^{i(kx-\omega t)}dk \qquad ...(1)$$

Now using $p = \hbar k$ and $E = \hbar\omega$. we get equation (1) as

$$\psi(x,t)=\frac{1}{\sqrt{2\pi\hbar}}\int_{\infty}^{\infty} a(p)\exp\left[\frac{i}{\hbar}(px-Et)\right]dp$$

Thus

$$\psi(x,\ 0)=\frac{1}{\sqrt{2\pi\hbar}}\int_{\infty}^{\infty} a\,(p)\ \exp\left(\frac{i}{\hbar}px\right)dp$$

$\frac{1}{\sqrt{2\pi\hbar}}$ has been introduced so that *a(p)* is given by symmetrical relation

$$a\,(p)=\frac{1}{\sqrt{2\pi\hbar}}\int_{\infty}^{\infty} \psi(x,\ 0)\ \exp\left[-\frac{i}{\hbar}px\right]dx$$

- ***Group velocity***

$$v_g=\frac{d\omega}{dk}=\frac{dE}{dp}\left(\text{U}\sin g\ \begin{matrix}\bar{p}=\hbar k\\ \bar{E}=\hbar\omega\end{matrix}\right)$$

- ***Phase velocity***

$$v_p=\frac{\omega}{k}$$

The One Dimensional Gaussian Wave Packet: Propagation of Gaussian wave packet $\psi(x_1\ t = 0)$

$$=\frac{1}{\left(\pi\sigma^2\right)^{1/4}}\exp\left[-\frac{(x-x_0)^2}{2\sigma^2}\right]\exp\left[\frac{i}{\hbar}p_0x\right]$$

- Wave function is normalised
- $<x>=\int_{\infty}^{\infty}\psi^*x\,\psi\,dx=x_0$
- $<x^2>\ =\int_{\infty}^{\infty}\psi^*x^2\,\psi\,dx=\frac{\sigma^2}{2}+x_0{}^2$

- $$<p> = \int_{-\infty}^{\infty} \psi^* \left(-i\hbar \frac{\partial \psi}{\partial x} \right) dx = p_0$$

- $$<p^2> = \int_{-\infty}^{\infty} \psi^* \left(-\hbar^2 \frac{\partial^2}{\partial x^2} \psi \right) dx = \frac{\hbar^2}{2\sigma^2} + p_0^2$$

Simple Solutions of the Schrodinger Equation

I : One Dimensional Consideration

Orthogonality Condition

$$\int_{-\infty}^{\infty} \psi_n^* \psi_n \, dx = 0$$

Normalization condition

$$\int_{-\infty}^{\infty} |\psi_n|^2 \, dx = 1$$

Orthonormality condition

$$\int_{-\infty}^{\infty} \psi_k^* \psi_n \, dx = \delta_{kn}$$

$\delta_{kn} \rightarrow$ kronecker delta function

$$\delta_{kn} = 0, \; k \neq n = 1, \; k = n$$

Particle in a One-Dimensional Infinitely Deep Potential Well

The allowed energy levels are given by

$$E_n = \frac{\pi^2 n^2 \hbar^2}{2ma^2}, \quad n = 1, 2, 3$$

$$= \frac{n^2 h^2}{8ma^2}$$

The corresponding eigenfunctions are

$$\psi_n = \sqrt{\frac{2}{a}} \sin\left(\frac{n\pi}{a} x\right)$$

Particle in a One-Dimensional Potential Well of Finite Depth

Energy eigenvalues would given as

$$E = \frac{(2p)^2 \pi^2 \hbar^2}{2ma^2} \; ; p = 1, 2, 3 \text{ for } V \to \infty$$

For $V_0 >> E$

$$E = \frac{(2p+1)^2 \pi^2 \hbar^2}{2ma^2}$$

Linear Harmonic Oscillator

The potential energy for this

$$V(x) = \frac{1}{2} m\omega^2 x^2$$

The energy eigenvalues

$$E = \left(n + \frac{1}{2}\right)\hbar\omega, \; n = 0, 1, 2, ...$$

Zero point energy

$$E = \frac{\hbar\omega}{2}$$

The normalised harmonic oscillator wave functions are

$$\psi_n(x) = N_n \exp\left(-\frac{1}{2}\alpha^2 x^2\right) H_n(\alpha x)$$

where, $$N_n = \left[\frac{\alpha}{\sqrt{\pi}\, 2^n \, n!}\right]^{1/2}$$

and $$\alpha = \left(\frac{m\omega}{\hbar}\right)^{1/2}$$

and H_n (αx) → Hermite polynomial.

The Double Well Potential

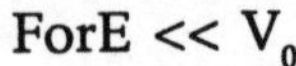

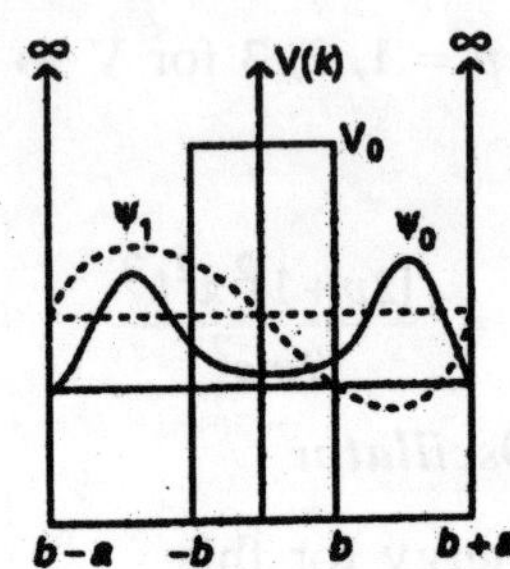

$$E_n^{(0)} = \frac{\pi^2 \hbar^2 n^2}{2ma^2}$$

and $$E_n = E_n^{(0)} - \frac{2E_n^{(0)}}{k_0 a} \mp 4\frac{E_n^{(0)}}{k_0 a} e^{-2k_0 b}$$

where $$k_0 = \left[\frac{2m}{\hbar^2}\left(V_0 - E_n^{(0)}\right)\right]^{1/2}$$

For $V \to \infty,\ E_n \to E_n^{(0)}$

- ***Wave function***

$$\psi(x, t) = \frac{1}{\sqrt{2}} e^{-iE_a^{(1)} t/k} \left[\psi_S^{(1)}(x) + \psi_a^{(1)}(x) e^{-i\pi t/r}\right]$$

where

$\psi_s^{(1)} \to$ Lowest symmetric wave functions

$\psi_a^{(1)} \to$ Lowest antisymmetric wave functions

One Dimensional Barrier Transmission Problems

The Potential Step

- The current densities associated with the wave function Ae^{ikx}, Be^{-ikx} and $ce^{ik_1 x}$ represent the incident current, reflected current and transmitted current respectively, these are

$$J_{inc} = \frac{\hbar k}{m}|A|^2$$

$$J_{ref} = \frac{\hbar k}{m}|B|^2$$

$$J_{tr} = \frac{\hbar k_1}{m}|C|^2$$

$$\Rightarrow \quad B = \frac{k-k_1}{k+k_2}A \text{ and } C = \frac{2k}{k+k_1}A$$

The reflection and transmission coefficients are

$$R = \frac{J_{ref}}{J_{inc}} = \frac{|B|^2}{|A|^2} = \frac{(k-k_1)^2}{(k+k_1)^2}$$

$$T = \frac{J_{tra}}{J_{inc}} = \frac{k_1|C|^2}{k|A|^2} = \frac{4kk_1}{(k+k_1)^2}$$

The Rectangular Potential Barrier

Transmission coefficient

$$T = \frac{-4{p_1}^2{p_2}^2 \operatorname{sec}h^2(ip_2\, a\, \hbar)}{\left({p_1}^2+{p_2}^2\right)\tanh^2(ip_2\, a\, \hbar) - 4{p_1}^2{p_2}^2}$$

$$R = \frac{\left({p_1}^2-{p_2}^2\right)^2\tanh^2(ip_1\, a\, \hbar)}{\left({p_1}^2+{p_2}^2\right)^2\tanh^2(ip_2\, a\, \hbar) - 4{p_1}^2{p_2}^2}$$

A special case when the barrier is thick

$$T = \frac{-16{p_1}^2{p_2}^2\, e^{2ip_2\, a/\hbar}}{\left({p_1}^2+{p_2}^2\right)^2 - 4{p_1}^2{p_2}^2} = \frac{-16{p_1}^2{p_2}^2\, e^{2ip_2\, a/\hbar}}{\left({p_1}^2-{p_2}^2\right)^2}$$

$$= \frac{+16E\,(V_0-E)}{{V_0}^2} e^{\left[-2\sqrt{\{2m(V_0-E)\}}a/\hbar\right]}$$

$$R = \frac{\left({p_1}^2-{p_2}^2\right)^2}{\left({p_1}^2-{p_2}^2\right)^2} = 1$$

Simple Solutions of the Schrodinger Equation

II: 3-Dimensional Consideration

The hydrogen atom. The Hamiltonian for hydrogen atom

$$H = -\frac{\hbar^2}{2m_1}\nabla_1^2 - \frac{\hbar^2}{2m_2}\nabla_2^2 + V\left(\left|\bar{r}_1 - \bar{r}_2\right|\right)$$

- The energy depends on the total quantum number *n*.
- For each *n* we have values of *l* ranging from 0 to *n* –*1*, and for each value of *l*, the *m* values range from –*l to* +*l* there are $\sum_{l=0}^{n-1}(2l+1) = n^2$, states belonging to a particular energy.

For the Ground State of hydrogen atom the Wave Function is Spherically Symmetric

$$\psi_{100} = \frac{1}{\sqrt{4}}\left(\frac{z}{a_0}\right)^{3/2} \exp\left(-\frac{zr}{a_0}\right)$$

P(*r*)*dr* → the probability of finding the electron between *r* and *r* + *dr* is

$$P(r)dr = |\psi_{100}|^2 4\pi r^2 dr$$

$$= 4\left(\frac{z}{a_0}\right)^3 r^2 \exp\left[-\frac{2zr}{a_0}\right] dr$$

The Particle in a Box (In 3-D)

- Wave function

$$\psi_{n_x n_y n_z} = \frac{2\sqrt{2}}{\sqrt{abc}} \sin\frac{n_x \pi x}{a} \sin\frac{n_y \pi y}{b} \sin\frac{n_z \pi z}{c}$$

- Energy levels

$$E_{n_x n_y n_z} = \frac{\hbar^2}{8m}\left[\frac{n_x^{\ 2}}{a^2} + \frac{n_y^{\ 2}}{b^2} + \frac{n_z^{\ 2}}{c^2}\right]$$

- Energy states between E and E + dE is

$$= 2\pi V\left(\frac{2m}{h^2}\right)^{3/2} E^{1/2}, \text{ V} \to \text{Volume}$$

Three-Dimensional Harmonic Oscillator

For isotropic oscillator, energy is given as

$$\text{E} = \left(n + \frac{3}{2}\right)\hbar\omega$$

- The degree of degeneracy is $\frac{1}{2}$ $(n + 1)$ $(n + 2)$

Angular Momentum

- Angular momentum operator representations of L_x, L_y, and L_z and $L^2 = L_x^{\ 2} + L_y^{\ 2} + L_z^{\ 2}$

$$L_x = -i\hbar\left(y\frac{\partial}{\partial z} - z\frac{\partial}{\partial y}\right)$$

$$L_y = -i\hbar\left(z\frac{\partial}{\partial x} - x\frac{\partial}{\partial z}\right)$$

$$Lz = -i\hbar\left(x\frac{\partial}{\partial y} - y\frac{\partial}{\partial x}\right)$$

- In terms of spherical polar coordinates

$$L_x = i\hbar\left(\sin\phi\frac{\partial}{\partial\theta} + \cot\theta\cos\phi\frac{\partial}{\partial\phi}\right)$$

$$L_y = i\hbar\left(-\cos\phi\frac{\partial}{\partial\theta} + \cot\theta\sin\phi\frac{\partial}{\partial\phi}\right)$$

and $\quad L_z = -i\hbar\frac{\partial}{\partial\theta}$

$$L_y = -\hbar^2 r^2 \left\{ \nabla^2 - \frac{1}{r^2}\frac{\partial}{\partial r}\left(r^2 \frac{\partial}{\partial r} \right) \right\}$$

Eigenvalues of L^2:

$$L^2 Y_{lm}(\theta, \phi) = l(l+1)\,\hbar^2 Y_{lm}(\theta, \phi)$$

eigenvalues of L^2 is $l(l+1)\hbar^2$

and for $L_z \rightarrow m\hbar$

The Commutation Relations

- $$[L_x, L_y] = i\hbar L_z;$$
  $$[L_y, L_z] = i\hbar L_x;$$
  $$[L_z, L_x] = i\hbar L_y$$
- $[L^2, L_x] = [L^2, L_y] = [L^2, L_z] = 0$

  Angular momentum in general

  $$J = L + Se$$
  $$J^2 = J_x^2 + J_y^2 + J_z^2$$

  Ladder operators :

  $$J_+ = J_x + iJ_y$$
  $$J_- = J_x - iJ_y$$

  Then $[J_z, J_+] = \hbar J_+$

  $$[J_z, J_-] = -\hbar J_-$$

  and $[J_+, J_-] = 2\hbar J_z$

  $$[J^2, J_+] = [J^2, J_-] = 0$$

Expectation Value

$$< J^2 > = < J_x^2 > + < J_y^2 > + m^2\hbar^2$$

J_x and J_y are Hermitian and the expectation value of Hermitian are non-negative.

Operator	*Eigenvalue*	*Eigenket*
J^2	$\lambda\hbar^2 = j(j+1)\hbar^2$	$\|jm>$
J^2	$\lambda\hbar^2 = j(j+1)\hbar^2$	$J \pm \|jm>$
J_z	$m\hbar$	$\|Jm>$
	$(m+1)\hbar$	$J_+ \|jm>$
	$(m-1)\hbar$	$J_- \|jm>$

$\Rightarrow$ $\qquad J_+ |jm> = |j, m+1>$

and $\qquad J_- |jm> = |j, m-1>$

Formulation of Matrix

- Matrix element for J^2 and J_z are given as

$$<j'm'\left|J^2\right|jm> \;=\; j(j+1)\hbar^2\,\delta_{jj'}\,\delta_{mm'}$$

$$<j'm'\left|J_z{}^2\right|jm> \;=\; m\hbar\,\delta_{jj'}\,\delta_{mm'}$$

where $<j'm'$ |and| $jm>$ represents column and row respectively.

Thus each matrices of J^2 and J_z has the form

j	m \ j', m'	$j'=0$, $m'=0$	$j'=\frac{1}{2}$, $m'=\frac{1}{2}$, $-\frac{1}{2}$	$j'=1$, $m'=1$, 0, -1 ………
0	0	(element present)	(0)	(0)……
$\frac{1}{2}$	$\frac{1}{2}$	(0)	(element present)	(0)……
$\frac{1}{2}$	$-\frac{1}{2}$			
1	1			
1	0	(0)	(0)	(element present) …
1	-1			

Note : J^2 and J_z matrices contain only principal diagonal

For Example :

1. $j = 1$

	j'	1	1	1
	m'	1	0	−1
j	m			
1	1	A	0	0
1	0	0	B	0
1	−1	0	0	C

$J^2 =$ (above table)

[order of matrix = $(2j + 1)$ = $(2 \cdot 1 + 1)$ = 3]

$$A = B = C$$

$$= j(j+1)\hbar^2 = 1(1+1)\hbar^2 = 2\hbar^2$$

$$\therefore \qquad J^2 = \begin{pmatrix} 2\hbar^2 & 0 & 0 \\ 0 & 2\hbar^2 & 0 \\ 0 & 0 & 2\hbar^2 \end{pmatrix} = 2\hbar^2 \begin{pmatrix} 1 & 0 & 0 \\ 0 & 1 & 0 \\ 0 & 0 & 1 \end{pmatrix}$$

and $\qquad J_z = m\hbar\, \delta_{jj'}\, \delta_{mm'}$ [As $m = 1, 0, -1$ for $j = 1$]

$$= \begin{pmatrix} 1.\hbar^2 & 0 & 0 \\ 0 & 0.\hbar^2 & 0 \\ 0 & 0 & -1.\hbar^2 \end{pmatrix}$$

$$= \hbar^2 \begin{pmatrix} 1 & 0 & 0 \\ 0 & 0 & 0 \\ 0 & 0 & -1 \end{pmatrix}$$

2. For $\qquad j = \dfrac{1}{2}$ [order of matrix = 2 × 2]

$$J^2 = \frac{3}{4}\hbar^2 \begin{pmatrix} 1 & 0 \\ 0 & 1 \end{pmatrix}$$

and $$J_z = \frac{1}{2}\hbar\begin{pmatrix}1 & 0\\ 0 & -1\end{pmatrix}$$

- Matrix elements of J_x and J_y in a representation in which J^2 and J_z are diagonal.

$$< j'm'|J_x|jm> = \frac{1}{2}\hbar\sqrt{\{(j-m)(j+m+1)\}}\,\delta_{jj'}\,\delta_{m',\,m+1}$$

and $$< j'm'|J_y|jm> = \frac{i\hbar}{2}\sqrt{\{(j+m)(j-m+1)\}}\,\delta_{jj'}\,\delta_{m',\,m-1}$$

- Matrix element for J_+, J_-

$$< j'm'|J_+|jm> = \hbar\sqrt{\{j(j+1)-m\,(m+1)\}}\,\delta_{jj'}\,\delta_{m',\,m+1}$$

and $$< j'm'|J_-|jm> = \hbar\sqrt{\{j(j+1)-m\,(m-1)\}}\,\delta_{jj'}\,\delta_{m',\,m-1}$$

For example

1. $$j = \frac{1}{2}$$

$$J_+ = \hbar\begin{pmatrix}0 & 1\\ 0 & 0\end{pmatrix}$$

and $$J_- = \hbar\begin{pmatrix}0 & 0\\ 1 & 0\end{pmatrix}$$

2. $$j = 1$$

$$J_+ = \sqrt{2}\,\hbar\begin{pmatrix}0 & 1 & 0\\ 0 & 0 & 1\\ 0 & 0 & 0\end{pmatrix}$$

and $$J_- = \sqrt{2}\,\hbar\begin{pmatrix}0 & 0 & 0\\ 0 & 0 & 0\\ 0 & 1 & 0\end{pmatrix}$$

- Matrix elements for L+ and L–

$$<l'm'|L_+|lm> = \hbar\sqrt{\{l(l+1)-m(m+1)\}}\,\delta_{ll'}\,\delta_{m',m+1}$$

$$<l'm'|L_-|lm> = \hbar\sqrt{\{l(l+1)-m(m-1)\}}\,\delta_{ll'}\,\delta_{m',m-1}$$

For $j = 1$

$L_+ =$ l'	m' | m	1	0	−1
1	1	0	$\hbar\sqrt{2}$	0
1	0	0	0	$\hbar\sqrt{2}$
1	−1	0	0	0

$$= \hbar\begin{pmatrix} 0 & \sqrt{2} & 0 \\ 0 & 0 & \sqrt{2} \\ 0 & 0 & 0 \end{pmatrix}$$

Similarly, $$L_- = \begin{pmatrix} 0 & 0 & 0 \\ \sqrt{2} & 0 & 0 \\ 0 & \sqrt{2} & 0 \end{pmatrix}$$

Spin Angular Momentum

- A particle having intrinsic (spin) angular momentum 'S' which is unrelated to its orbital motion. Also it obey

$$[S, L] = 0$$

- $$S^2\,|SM_S> = s(s+1)\,\hbar^2\,|SM_S>$$

$$S_z\,|SM_S> = M_s\hbar^2\,|SM_S>$$

where $M_S = S, S-1, \ldots -s$, there are $(2s+1)$ values of M_S for given S and $(2s+1)$ states are possible

- For $s = \frac{1}{2}; M_s = \frac{1}{2}, -\frac{1}{2}$

Only $2\left(\frac{1}{2}\right)+1 = 2$ states are possible $|s,M_s>$

$\left|\frac{1}{2}, \frac{1}{2}\right> \uparrow$ = α state $\qquad \left|\frac{1}{2}, -\frac{1}{2}\right> \downarrow$ = β state

$$\therefore \quad \alpha = \begin{pmatrix} 1 \\ 0 \end{pmatrix} \uparrow \text{ and } \beta = \begin{pmatrix} 0 \\ 1 \end{pmatrix} \downarrow$$

α is wave function for $m_s = \frac{1}{2}$ or spin up while β stands for $m_s = -\frac{1}{2}$ or spin down wave function.

- General wave function

$$|x> \; = \begin{pmatrix} a \\ b \end{pmatrix} = a\begin{pmatrix} 1 \\ 0 \end{pmatrix} + b\begin{pmatrix} 0 \\ 1 \end{pmatrix} = a\alpha + b\beta$$

a →Probability amplitude that particle has spin up and

b →Probability amplitude for particle has spin down

- Also $S = \frac{1}{2}\hbar\sigma$.

σ →Pauli matrices and defined as

$$\sigma_x = \begin{pmatrix} 0 & 1 \\ 1 & 0 \end{pmatrix}; \; \sigma_y = \begin{pmatrix} 0 & -i \\ i & 0 \end{pmatrix}; \; \sigma_z = \begin{pmatrix} 1 & 0 \\ 0 & -1 \end{pmatrix}$$

- $S_+ = S_x + iS_y$

$S_- = S_x - iS_y$

or,

$$S_+ = \frac{1}{2}\hbar\sigma_+ = \frac{\hbar}{2}\left[\sigma_x + i\sigma_y\right]$$

$$= \frac{\hbar}{2}\left[\begin{pmatrix} 0 & 1 \\ 1 & 0 \end{pmatrix} + \begin{pmatrix} 0 & 1 \\ -1 & 0 \end{pmatrix}\right]$$

$$= \frac{\hbar}{2}\begin{pmatrix} 0 & 2 \\ 0 & 0 \end{pmatrix} = \hbar\begin{pmatrix} 0 & 1 \\ 0 & 0 \end{pmatrix} \text{ Raising operator}$$

$$S_- = \frac{1}{2}\hbar\sigma_- = \hbar\begin{pmatrix} 0 & 0 \\ 1 & 0 \end{pmatrix} \text{ Lowering operator}$$

Properties of Pauli Spin Matrix

(1) $\sigma_x^2 = \sigma_y^2 = \sigma_z^2 = 1$ and $\sigma^3 = 3$

(2) $\sigma_x \sigma_y = -\sigma_y \sigma_x = i\sigma_z$

(3) $\sigma_+^2 = \sigma_-^2 = 0$

(4) $\sigma_+ \sigma_- = 2(1+\sigma_z)$

(5) $(\bar{\sigma}.\bar{A})(\bar{\sigma}.\bar{B}) = (\bar{A}.\bar{B}) + i\,\bar{\sigma}.(\bar{A}\times\bar{B})$

Addition of Angular Momentum

- j_1 and j_2 are two independent angular momentum. The eigenvector $\psi\ (j_1 j_2 . m_1 m_2)$ defines eigenstates with four observable respectively $j_1\ (j_1 + 1)$, m_1, $j_2\ (j_2 + 1)$ and m_2.
- For given values of j_1 and j_2 *there* will be $(2j_1 + 1)$ possible values of m_1 and $(2j_2 + 1)$ possible values of m_2.

$\therefore$ There will be $(2j_1 + 1)\ (2j_2 + 1)$ eigenstates.

Clebsch-Gorden Coefficients (C-G Coeff.) Addition of Angular Momenta

$$\left.\begin{aligned}&\sum_{jm} \langle m_1 m_2 | JM \rangle \langle JM | m_1' m_2' \rangle \\ &\qquad = \delta_{m_1 m_1'}\,\delta_{m_2 m_2'} \\ &\sum_{m_1 m_2} \langle JM | m_1 m_2 \rangle \langle m_1 m_2 | J' M' \rangle \\ &\qquad = \delta_{JJ'}\,\delta_{MM'}\end{aligned}\right] \text{Definition}$$

Recurrence Relations for Calculation of C. G. Coefficient

(1) $$[J(J+1) - m(m+1)]^{1/2} \langle m_1 m_2 \mid JM+1 \rangle$$
$$= [j_1(j_1+1) - m_1(m_1-1)]^{1/2} \langle m_1 - 1, m_2 \mid JM \rangle + [j_2(j_2+1) - m_2(m_2-1)]^{1/2} \langle m_1, m_2 - 1 \mid JM \rangle$$

(2) $[J(J+1)-m(m-1)]^{1/2} < m_1 m_2 \mid Jm-1 >$

$$= [j_1(j_1+1)-m_1(m_1+1)]^{1/2} < m_1+1,$$

$$m_2 \mid JM > + [j_2(j_2+1)-m_2(m_2+1)]^{1/2}$$

$$< m_2+1, m_1 \mid JM >$$

So to determine C.G. coefficient $< m_1 m_2 \mid JM >$ we use recurrence relation (1) and (2).

Operator Method in Quantum Mechanics

If the operator $\hat{\alpha}$, $\hat{\beta}$ and $\hat{\gamma}$ are of such that

$$\hat{\gamma} f(x) = \hat{\alpha} f(x) + \hat{\beta} f(x)$$

- The operator $(\hat{\alpha}\hat{\beta} - \hat{\beta}\hat{\alpha})$ is called the commutator of the operator $\hat{\alpha}$ and $\hat{\beta}$.

e.g., $\quad \hat{x}\,\hat{p}\,\psi(x) - \hat{p}\,\hat{x}\,\psi(x) = -\frac{\hbar}{i}\psi(x)$

$\therefore \quad \hat{x}\,\hat{p} - \hat{x}\,\hat{p} = i\hbar$

Theorem : The necessary and sufficient condition that simultaneous eigenfunctions exist for two operators $\hat{O}_1$ and $\hat{O}_2$ is that they commute.

- Momentum operator $\rightarrow -i\hbar\frac{d}{dx}$

and a plane wave which represent a state of definite momentum p

$$\psi = \frac{1}{\sqrt{2\pi\hbar}} e^{(1/\hbar)\,p_x}$$

$\rightarrow$ Normalised eigenfunction

- If an operand u is a simultaneous eigenfunction of two operators $\hat{\alpha}$ and $\hat{\beta}$ such that

$$\hat{\alpha}\, u = \lambda u \text{ and } \hat{\beta} = \mu u$$

where λ and μ are numerical constants, then u is also an eigenfunction of the operator $\left(\hat{\alpha}\,\hat{\beta}-\hat{\beta}\,\hat{\alpha}\right)$ and belongs to the eigenvalue 0.

⇒ Operator associated with linear momentum

$$\bar{p} = -i\hbar\nabla$$

Energy operator

$$\bar{E} = i\hbar\frac{\partial}{\partial t}$$

and velocity operator,

$$\bar{v} = -\frac{i\hbar}{m}\nabla$$

Hermitian Operator

For any two well behaved functions $f(x)$ and $g(x)$ vanishing at infinity, an operator $\hat{O}$ satisfying the equation

$$\int_{-\infty}^{\infty} g^*(x)\left[\hat{O} f(x)\right]dx = \int_{-\infty}^{+\infty}\left[\hat{O}g(x)\right]^* f(x)\,dx$$

is said to be Hermitian.

- ***Parity Operator*** π defined as

$$\hat{\pi}\,\psi(x) = \psi(-x)$$

Eigenvalue ⇒ ± 1

+1 → Corresponds to even eigenfunction.

–1 → Corresponds to odd eigenfunction.

$\Rightarrow$ $\hat{\pi}$ operator commutes with Hamiltonian operator $\hat{H}$

- ***Projection Operator***

$$\hat{p}i = |\psi_i><\psi_i| \rightarrow \text{Projection operator}$$

$$\sum_{i=1}^{\infty} \hat{p}_i = 1$$

Properties

(1) $\hat{p}_i^2 = \hat{p}_i$

(2) $\hat{p}_i\hat{p}_j = \hat{p}_j\,\delta_{ij}$

(3) Projection operator $\hat{p}_i$ is Hermitian

Unitary Operator : When the inverse and adjoint of an operator all identical, then the operator is known as unitary operator. If $\hat{U}$ is unitary operator, then

$$\hat{U}^{+} = \hat{U}^{-1}$$

Properties of Hermitian Operations

(a) The sum of two Hermitian operators is Hermitian.

(b) The product of the Hermitian operators is Hermitian if and only if they commute.

(c) The operators as position $\hat{r}$, momentum

$$\hat{p} = \left(\frac{\hbar}{i}\right)\nabla \text{ and energy } \hat{H} = \left(\frac{p^2}{2\mu}\right) + V(r) \text{ are Hermitian.}$$

(d) Hermitian operators give real eigenvalues.

Theorem : The eigenfunctions of a Hermitian operator

corresponding to different eigenvalues will be orthogonal function.

Equation of Motion

$$\dot{f} = [f, H]$$

or,
$$\dot{x} = [x, H]$$

Theory of Scattering

The complete wave function in this case

$$\psi \underset{r\to\infty}{\longrightarrow} e^{i\underline{k}.\underline{r}.} + f(\theta)\frac{e^{ikr}}{r}.$$

$e^{ikr} \rightarrow$ incident wave and $f(\theta)\frac{e^{ikr}}{r}$ Scattered wave

where $f(\theta) \rightarrow$ Scattering amplitude

Differential scattering cross section

$$\sigma(\theta) = |f(\theta)|^2$$

Total scattering cross section (for spherically Symmetrical potential)

$$\sigma = 2\pi \int_0^{+\infty} \sigma(\theta) \sin\theta \, d\theta$$

The Born Approximation

Scattering amplitude

$$f(\theta, \phi) = -\frac{2\mu}{\hbar^2} \int V(r)\, e^{i\underline{k}.\underline{r}}\, d\underline{r}$$

If potential is spherically symmetric

$$f(\theta) = -\frac{1}{K}\int_0^{\infty} r \sin Kr\, V(r)\, dr$$

$$K = 2k \sin\frac{\theta}{2} \qquad k \rightarrow \text{propagation}$$

constant

The differential scattering cross section

$$\sigma(\theta) = |f(\theta)|^2 = \left|-\frac{1}{K}\int_0^\infty r \sin Kr\, V(r)\, dr\right|^2$$

Condition for validity of Born approximation

$$\left|\frac{1}{4\pi}\int_0^\infty \frac{e^{ik|r-r'|}}{|\underline{r}-\underline{r}'|}\cup(r')\, e^{i\underline{k}.\underline{r}'}\, dr'\right| << 1$$

The Method of Partial Waves : The method of partial waves is mainly applicable to spherically symmetric potential and consists of the expansion of the wave function as a series of spherical harmonics multiplied by a radial wave function.

Scattering amplitude

$$f(\theta) = \sum_{l=0}^{\infty} \frac{(2l+1)}{k} \sin\, \delta_l\, e^{i\delta_l}\, P(\cos\,\theta)$$

The D.C.S. (Differential Scattering cross section)

$$\sigma(\theta) = \sum_{l=0}^{\infty} \frac{(2l+1)^2 P_l^{\,2}(\cos\,\theta)\, \sin^2\,\delta_l}{k^2}$$

The T.C.S. (Total scattering cross section)

$$\sigma = \frac{4\pi}{k^2}\sum_{l=0}^{\infty} (2l+1)^2 \sin^2\,\delta_l$$

Optical Theorem

$$\sigma_t = \frac{4\pi}{k}.\text{Im}\, f(0)$$

where $\quad \text{Im}\, f(0) = \dfrac{1}{k}\displaystyle\sum_{l=0}^{\infty} (2l+1)^2 \sin^2\,\delta_l$

Scattering by Hard Sphere : For sufficiently low energies

TCS $\sigma \approx 4\pi a^2$

For high energies

$$\sigma \approx 2\pi a^2$$

Perturbation Theory

Hamiltonian for He[$e^- - e^-$ interaction]

$$H = \sum_{i=1}^{2}\left(-\frac{\hbar}{2\mu}\nabla_i^2, -\frac{ze^2}{r_i}\right)+\frac{e^2}{r_{12}}$$

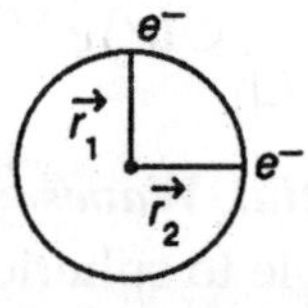

$\frac{e^2}{r_{12}} \rightarrow$ Produces problem in solving the problem exactly.

This term is known as perturbation.

Time-Independent Perturbation Theory

- When a small disturbance is applied we can break Hamiltonian as

$$\text{H} = \text{H}^\circ + \text{H}'$$

H° → Non-perturbed, term

H′ → Small perturbed term

(i) ***Non-degenerate Case :*** *(i.e.,* there will be one eigenfunction corresponding to each (energy level) eigenvalue)

In 1st Order Perturbation

1st order corrections are

(a) $\text{E} = E_m + \langle m|H'|m\rangle$

$$\langle m|H'|m\rangle$$

1st order correction in energy

(b) $$|\psi>^{(1)} = |m> + \underset{k}{S'} \frac{<k|H'|m>}{(E_m - E_k)} |k>$$

where $$|\psi_1> \;=\; \underset{k}{S'} \frac{<k|H'|m>}{(E_m - E_k)} |k>$$

1st order correction in wave function.

IInd Order Perturbation

$$W_2 \;=\; \underset{k}{S'} \frac{[<m|H'|k>]^2}{(E_m - E_k)}$$

This is IInd order correction in energy.

(ii) *Degenerate Case :* (Two or more eigenvalues corresponding to one eigenfunction).

1st order correction in energy

$$W_1 = \frac{(H'_{mm} + H'_{ll}) \pm \sqrt{(H'_{mm} - H'_{ll})^2 + 4H'_{ml}H'_{lm}}}{2}$$

Degeneracy is removed in 1st order.

where

$$H'_{mm} \;=\; <m|H'|m>$$

$$H'_{ll} \;=\; <l|H'|l>$$

$$H'_{ml} \;=\; <m|H'|l>$$

$$H'_{lm} \;=\; <l|H'|m>$$

1st Order Correction in Wave Function

$$|\psi_1> \;= -\underset{k}{S'} \frac{[H'_{km} a_m + H'_{kl} a_l]}{(E_k - E_m)} |k>$$

IInd Order Correction in Energy

$$W_2 = \frac{1}{2} \underset{n}{S'} \frac{(H'^2{}_{mm} + H'^2{}_{ln})}{(E_m - E_n)}$$

$$\pm\frac{1}{2}\left[\underset{n}{S'}\left(\frac{H'^2{}_{mn}+H'_{\ln}}{(E_m-E_n)}\right)^2+\frac{4H'_{mn}H'_{\ln}H'_{nm}H'_{nl}}{(E_m-E_n)^2}\right]^{1/2}$$

Degeneracy is removed in IInd order.

Zeeman Effect (Without e-Spin)

The effect of uniform external magnetic field on energy levels of an atom, is called the "Zeeman effect".

- Perturbation

$H'=\frac{e}{2\mu}(\vec{B}\,.\,\vec{L})$ represents the energy $(-\vec{M}\,.\,\vec{B})$ of a magnetic dipole with $\vec{M}=\frac{-e}{2\mu}\vec{L}$

So 1st order energy correction term

$$E_1 = \left\langle m\left|\frac{e}{2\mu}BL_z\right|m\right\rangle = \frac{eB}{2\mu}m\hbar$$

So, for a given *n*, *m* can take value from $-l$ to l *i.e.,* the degeneracy is of $(2l+1)$ states. So, energy levels are splited by applying magnetic field. The nth level being splited into $(2l+1)$ levels. This is Zeeman effect.

Stark Effect : The splitting of energy levels of an atom caused by a uniform external electric field $\vec{E}$ is called Stark effect.

- For ground state there is no 1st order Stark effect.
- For 1st excited state, $n = 2$

There is four fold degeneracy. For which we have four roots 0, 0, $3e\,Ea_0$ and $-3eEa_0$. So, two roots 0, 0 are same and hence two fold degeneracy have been removed by 1st order Stark effect.

So, behaviour of hydrogen atom in 1st excited state $(n - 2)$ is like a permanent electric dipole moment of, magnitude $3ea_0$ that can be mapped in three different ways.

(1) One state parallel to external field.

(2) One state antiparallel to field.

(3) Two states with zero component along field.

Original level —— $\updownarrow 3eEa_0$ / $\updownarrow 3eEa_0$ Splitted level

Fermi's Law

Transition probability per unit time is

$$T = \frac{2\pi}{\hbar}\left|H_{ml}^{(1)}\right|^2 \rho(m)$$

$$T = \frac{2\pi}{\hbar}\left|H_{ml}^{(1)}\right|^2 \rho(m)$$

$$H_{ml}^{(1)} = < m\left|H^{(1)}\right| l >$$

This is Fermi's Golden rule and states that the transition probability per unit time

(i) Is non-zero only between continuum states of the same energy.

(ii) Is proportional to the square of $\left|H_{ml}^{(1)}\right|$ of the perturbation connecting the states.

(iii) Is proportional to the density of final states.

- The probability of finding the system in mth state—

Considering the mth level above lth level, then

$\omega_{ml} = (E_n - E_l)/\hbar$ = positive quantity

If lth level above mth level, then

$\omega_{ml} = (E_n - E_l)/\hbar$ = negative quantity

So, when $E_m > E_l$, the probability is

$$= \frac{4\left|H^{\circ(1)}_{ml}\right|^2}{\hbar^2}\left[\frac{\sin^2(\omega_{ml}-\omega)t_0/2}{(\omega_{ml}-\omega)^2}\right]$$

When $El > E_{m'}$, then the probability is

$$= \frac{4\left|H^{o(1)}_{ml}\right|^2}{\hbar^2}\left[\frac{\sin^2(\omega_{ml}-\omega)t_0/2}{(\omega_{ml}-\omega)^2}\right]$$

Identical Particles and Spin

Postulates

(1) The identical particles having an integral quantum number for their intrinsic spins are described by the symmetric wave function of type

$\psi(1, 2, 3,..., r,... s,..., N) = +\psi(1, 2, 3,...s,...r,...N)$

The particles described by this type of wave function are known as Bose particles (Bosons). Examples of Bosons are photon (spin 1), neutral helium atom in their normal state (spin 0) and α-particle (spin 0) etc.

(2) The particles with half odd integral quantum number for their intrinsic spins are described by antisymmetric wave function and given as

$\psi(1, 2, 3,..., r,... s,..., N) = -\psi(1, 2, 3,...s,...r,...N)$

This type of particle are known as Fermi particle (Fermions) because they obey Fermi-Dirac statistics.

Examples : Electrons, protons, neutrons, muons (all with spin 1/2)

- Antisymmetric wave function can be built by adding all the functions with even number of interchanges and subtracting the sum of all those with an odd number of interchanges.

Pauli's Exclusion Principle

- The probability of two antisymmetric identical particles is zero or two non-interacting Fermions cannot both be in same quantum state.

(11)

Structural Crystals

Crystal structure is explored through the diffraction of waves having a Wavelength comparable with the interatomic spacing (10^{-10} m) in crystals. Radiation of longer wavelength cannot resolve the details of structure, while radiation of much shorter wavelength is diffracted through inconveniently small angles. Usually, diffractions of X-rays, neutrons and, less often, electrons is employed in the study of crystal structure.

W. L. Bragg presented an explanation of the observed diffracted beams from a crystal. He supposed that the incident waves undergo specular (mirror-like) reflection at the *various* parallel planes of atoms in the crystal, with each plane reflecting only a small fraction of the radiation. *The diffracted beams are found only when the reflections from the various planes of atoms interfere constructively.*

In the below given figure is shown a particular set of atomic planes in a crystal, *d* being the inter-planar spacing. Suppose an X-ray beam is incident at a glancing angle θ. It is scattered by the atoms like *A* and *B* in random directions. Constructive interference takes place only between those scattered waves which are reflected specularly and have a path difference of $n\lambda$, where λ is the X-ray wavelength and n is an integer. The path difference for the waves reflected from adjacent planes is

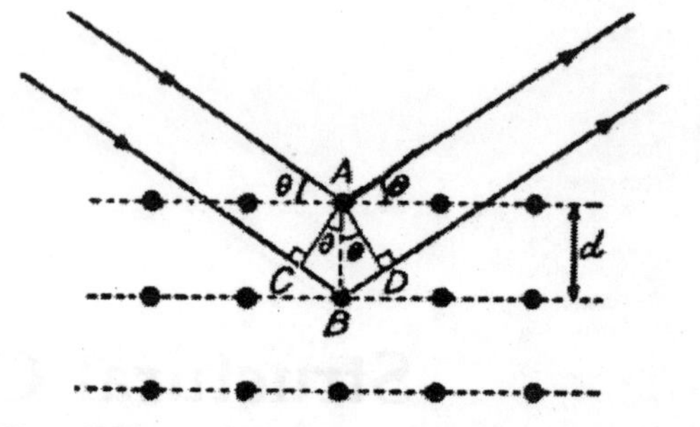

$$CB + BD = d \sin \theta + d \sin \theta = 2d \sin \theta.$$

For constructive interference, we must have

$$\boxed{2d \sin \theta = n\lambda.} \quad n = 1, 2, 3, \ldots\ldots$$

This is Bragg's law. It shows that for given values of λ and d, only for certain values of θ (corresponding to n = l, 2, 3,......), the reflected waves add up in phase to give a strong reflected (diffracted) beam. The images obtained in these directions are the first, second, third.......order diffraction images.

The Bragg law is a consequence of the periodicity of the space lattice. The arrangement of atoms (composition of the basis) associated with each lattice point determines the relative intensity of the various orders n of diffraction from a given set of parallel planes.

The maximum value of sin θ is 1. Therefore, Bragg reflection can occur only for wavelengths $\lambda \leq 2d$, and d is of the order of 1 Å. This is why we cannot use visible light.

The Bragg law requires that θ and λ should be so related as to satisfy the equation $2d \sin \theta = n\lambda$. X-rays of wavelength A striking a three-dimensional crystal at any angle of incidence will, in general, not be reflected. To satisfy the Bragg law in experiment, either λ or θ is varied.

There are four experimental methods of diffraction used in crystal structure analysis : (*i*) Bragg method, (*ii*) Laue method, (*iii*) rotating crystal method and (*iv*) Powder method.

Crystal Structure by Bragg

The Bragg's method of studying the crystal structure consists in finding out the interplanar spacings, d_{hkl}, of the various sets of planes rich in atoms. From a knowledge of d_{hkl}, and from the relative intensities of reflection for different orders, the crystal structure can be deduced.

The experimental arrangement required is a Bragg's X-ray spectrometer (Fig.). The crystal *C* under investigation is mounted on the spectrometer table so that its (100) face is on the axis of rotation. The position of the table can be read on a circular scale provided with a vernier V_1. A strong, *monochromatic* X-ray beam is first passed through two fine slits S_1 and S_2 and then made to fall on the crystal face. The reflected beam after passing through the silt S_3 enters an ionisation chamber *D* mounted on an arm which can be rotated about the same axis as the crystal. Its position can be read by a second vernier V_2. The gas in the chamber is ionised by the X-rays. The resulting ionising current, measured by the electrometer *E*, is a measure of the intensity of X-rays reflected by the crystal.

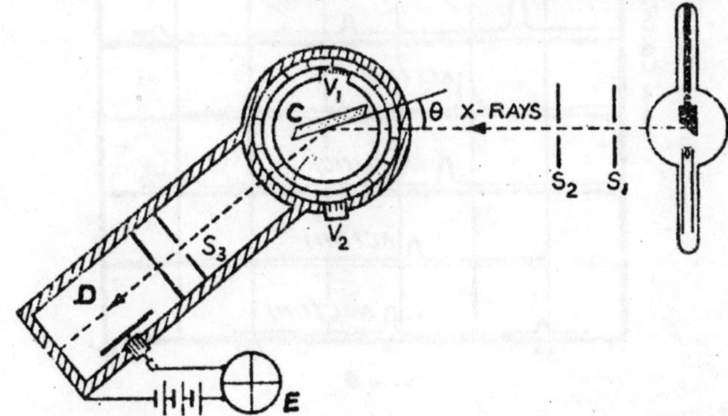

The crystal is rotated through small angles (while the arm carrying the ionisation chamber is rotated through double the angles) and the ionisation current is measured each time. The current is plotted against the glancing angle θ. The curve so obtained shows a number of peaks which correspond to those glancing angles θ which satisfy the Bragg's equation $2d_{100} \sin \theta = n\lambda$ (Fig.). The different peaks correspond to different orders (n = 1, 2, 3,...). Hence the corresponding glancing angles can be directly read from the curve. Now, the (110) face and the (111) face of the crystal are presented turn by turn to the incident beam of X-rays and the same procedure is gone through. Since today the absolute wavelength λ of a given X-radiation is known from measurements with ruled gradings, the interplanar spacings d_{100}, d_{110} and d_{111} can be determined from the Bragg's equation. If, however, the X-ray wavelength is not known, the ratios of the interplanar spacings can be determined.

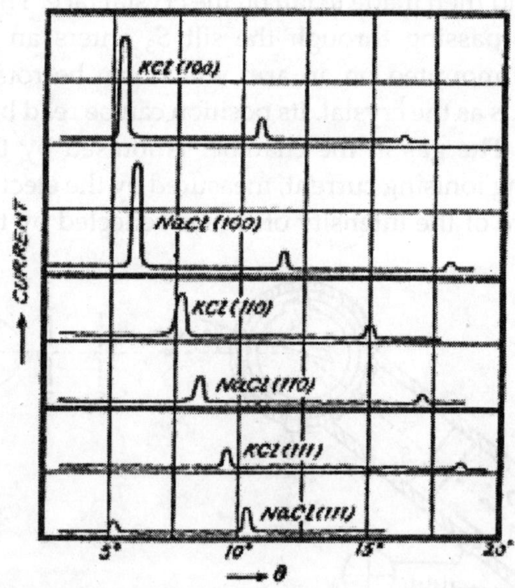

Suppose for a particular crystal used on a Bragg's spectrometer, strong reflections from the sets of planes (100);

(110); (111) are obtained for angles $\theta_1, \theta_2, \theta_3$ respectively in the first (n = 1) order. Then, from Bragg's equation, we have

$$2\, d_{100} \sin \theta_1 = \lambda\,,$$
$$2d_{110} \sin \theta_2 = \lambda\,,$$
and
$$2d_{110} \sin \theta_3 = \lambda.$$

Hence $\quad d_{100} : d_{110} : d_{111} = 1/\sin \theta_1 : 1/\sin \theta_2 : 1/\sin \theta_3.$

Bragg found that for a KCl crystal ; $\theta_1 = 5°23'$; $\theta_2 = 7°37'$ and $\theta_3 = 9°25'$. Hence for KCl

$$\begin{aligned} d_{100} : d_{110} : d_{111} &= 1/\sin 5°23' : 1/\sin 7°37' : 1/\sin 9°25' \\ &= 1/0.0938 : 1/0.1326 : 1/0.1636 \\ &= 1/1 : 1/1.414 : 1/1.744 \\ &= 1 : 1/\sqrt{2} : 1/\sqrt{3}\,. \end{aligned}$$

Theoretically, this ratio is found to hold for a simple cubic lattice structure. Hence it is concluded that the KCl *crystal has a simple cubic structure.* However, from these data we cannot know the constituents are the lattice corners, whether they are molecules of KCl or atoms (or ions) of K and Cl alternately. This can be decided from a study of other similar crystals such as NaCl.

When the first order reflections from the three faces (100), (110) and (111) of NaCl are compared, the ratios between the interplanar spacings are found as :

$$d_{100} : d_{110} : d_{111} = 1 : 1/\sqrt{2} : 2/\sqrt{3}\,.$$

which agree with the theoretical values for a faced-centered cubic lattice. From this we conclude that the NaCl *crystal has fee lattice* (whereas KCl had shown a simple cubic structure).

There is no reason why should there be a difference in the lattice structures of KCl and NaCI. The observed dissimilarity can be explained if we compare the curves of the two crystals. The curves for the (100) and (110) planes of the two crystals are similar in the sense that the intensity falls off continuously as we go to higher orders of reflection. However, *the curves for the* (111) *planes of the two crystals differ.* In case of NaCI (111) the second order is *stronger* than the first order and similarly, as Bragg found, the fourth order is stronger than the third. Such an alternation of intensity in the reflection from the (111) planes is observed in all alkali halides with the exception of KCl.

Bragg showed that the only way of explaining the intensity alternation is to assume that *the lattice points of the cubic structure are occupied by single ions and not by molecules,* and the ions of the two kinds (Na^+ and Cl^- or K+ and Cl^-) occur alternately, as shown in the figure below. Now it can be seen that the (100) and (110) planes contain equal number of ions of both the kinds. But the (111) planes contain, alternately, all Na^+ (or K^+) ions and all Cl^- ions. That is, the Na^+ (or K^+) planes occur exactly half-way between Cl^- planes. When the glancing angle is correct for the *first order* reflection from the Na^+ (or K^+) planes and also from the Cl^- planes, the rays reflected from a Na^+ (or K^+) plane are exactly out of phase with those reflected from the adjacent Cl^- plane thus causing destructive interference. Therefore, the first order reflection is weakened. For the second order reflection there will be reinforcement between all the reflected rays as seen by substituting in Bragg's equation. In general, for reflection at the (111) face the odd orders will be weaker compared with the even orders *i.e.* there will be an intensity alternation.

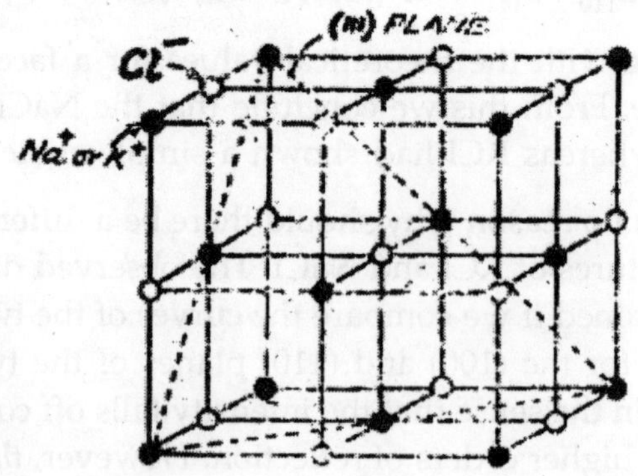

The fact that the intensity alternation is observed in case of NaCl (and other alkali halides) but not in KCl is readily explained when we realise that the reflecting (or scattering) powers of different ionic planes in a crystal are different. In fact, the reflecting power of an atom or ion is proportional to the number of electrons in it *i.e.* to the atomic number Z. Therefore, the Na^+ ions (Z = 10) and Cl^- ions (Z = 18) have quite different reflecting powers so that the rays reflected from the Na^+ and Cl^- planes appreciably differ in intensity and hence the destruction of the odd orders is not complete. That is, *in* NaCl *the odd orders become only weaker* but are still present. On the other hand, the K^+ ions (Z = 18) and Cl^- ions (Z = 18) have equal reflecting powers so that the rays reflected from the K^+ and Cl^- planes have equal intensities and the destruction of the odd orders is complete. That is, in *case of KCl crystal, the odd orders are entirely missing* and the intensity alternation is not observed. Thus in KCl (111) the first observed peak (Fig.), in fact, corresponds to the *second order* reflection (and not to the first order). When this correction is made, we get the same interplanar spacing ratios for KCl as for NaCl *i.e.*

$$d_{100} : d_{110} : d_{111} = 1 : 1/\sqrt{2} : 2/\sqrt{3}\,.$$

which correspond to an *fcc* lattice. The Na^+ (or K^+) ions alone form an *fcc* lattice, and so do the Cl^- ions. If the reflecting powers of the two kinds of ions are the same, the distinction between them disappears and the structure reduces to a simple cubic lattice as was first found in case of KCl.

Once the arrangement of atoms (or ions) in the, cubic crystal has been decided, we can calculate the absolute value of interplanar spacing.

Technique of Laue

In Laue's method, a single crystal is held *stationary* in a *continuous* X-ray beam. The crystal diffracts the discrete values of λ for which the crystal planes of spacing *d* and the incidence angle θ satisfy the Bragg's law.

The experimental arrangement is shown in Fig. (*a*). A continuous (≈ 0.2 Å to 2 Å) *X*-ray beam, well-collimated by a pinhole arrangement, is allowed to fall on the crystal. The crystal holder is adjusted to obtain proper orientation of the crystal. A flat film *A* is placed to receive the transmitted-diffracted beams (or a flat film *B* to receive the reflected-diffracted beams). The diffraction pattern is a regular one, consisting of a series of spots and is characteristic of crystal structure. It is called the Laue pattern (Fig *b*).

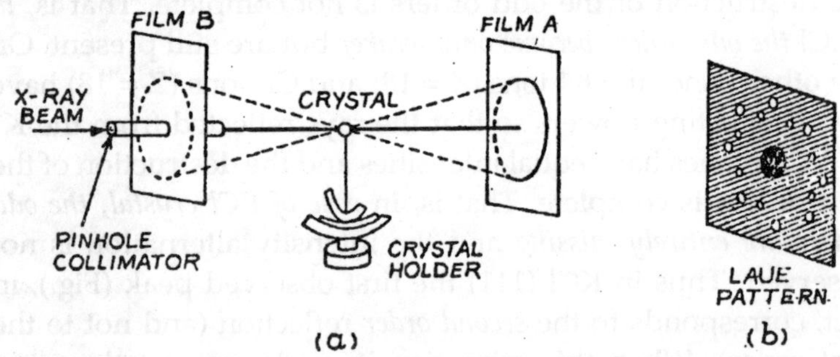

Each spot in the Laue pattern corresponds to an interference maximum for a set of crystal planes satisfying the Bragg equation $2d \sin \theta = n\lambda$ for a particular wavelength selected from the (continuous) incident beam. The atomic arrangement in the crystal can, therefore, be deduced from a study of the positions and intensities of the Laue spots.

The distribution of spots in the Laue pattern depends on the symmetry of the crystal and its orientation with respect to the *X*-ray beam. For example, if a crystal with four-fold axial symmetry is oriented with the axis parallel to the beam, then, the Laue pattern will show the four-fold symmetry, as in Fig. *b*. The Laue method is convenient for the rapid determination of crystal orientation and symmetry which is needed for solid state experiments. It is also used to study crystalline imperfections under mechanical and thermal treatment. The method is, however, not used for actual crystal structure determination. This is because it is possible for several wavelengths to reflect in different orders from the same plane, so that different orders may superpose on a single spot. This would make the analysis difficult.

Technique of Rotating Crystal

In this method, a single crystal is rotated about a fixed axis in a *monochromatic* beam of X-rays (or neutrons). The variation in the angle θ presents different atomic planes of the crystal to the X-ray beam for reflection.

A simple experimental arrangement is shown in Fig. below. The crystal under investigation is mounted on a rotating spindle, and the film is wrapped on a cylindrical surface coaxial with the spindle. The incident X-ray beam, made monochromatic by a filter or by reflection from an earlier crystal, is diffracted from a given crystal plane whenever in the course of rotation the value of *6* satisfies the Bragg equation. The beams reflected from all planes parallel to the (vertical) axis of rotation lie in the horizontal plane, while those from planes with other orientations lie above and below the horizontal plane.

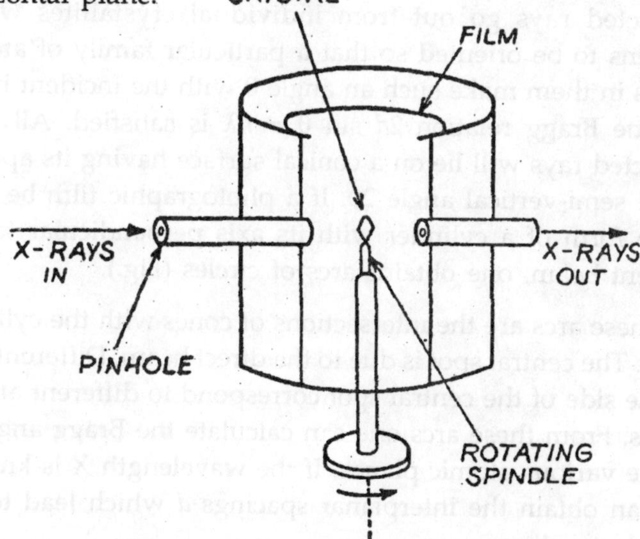

Modern rotating-crystal diffractometers use scintillation counters or proportional counters to detect the diffraction radiation. They automatically collect all the data regarding crystal structure.

Powder Method and Crystal Analysis

It is a more straight forward method for the analysis of crystal structure, developed independently by Debye and Scherrer in Germany and by Hull in America in 1916. In this method, the specimen, instead of being a single crystal, is a finely-powdered crystal.

The minute crystals, called crystallites, of this powder are all randomly (almost continuously) oriented so that they make all possible angles with the incident radiation. Thus all orders of reflection from all possible atomic planes are recorded at the same time.

The experimental arrangement is shown in Fig. below. A fine beam of monochromatic X-rays is allowed to fall on a thin-walled capillary tube *P* containing the powdered crystal. Diffracted rays go out from individual crystallites which happens to be oriented so that a particular family of atomic planes in them make such an angle θ with the incident beam that the Bragg relation $2d \sin \theta = n\lambda$ is satisfied. All such diffracted rays will lie on a conical surface having its apex at *P* and semi-vertical angle 2θ. If a photographic film be kept in the form of a cylinder with its axis perpendicular to the incident beam, one obtains arcs of circles (Fig.).

These arcs are the intersections of cones with the cylinder (film). The central spot is due to the direct beam. Different arcs on one side of the central spot correspond to different atomic planes. From these arcs one can calculate the Bragg angles θ for the various atomic planes. If the wavelength X is known, one can obtain the interplanar spacings *d* which lead to the crystal structure.

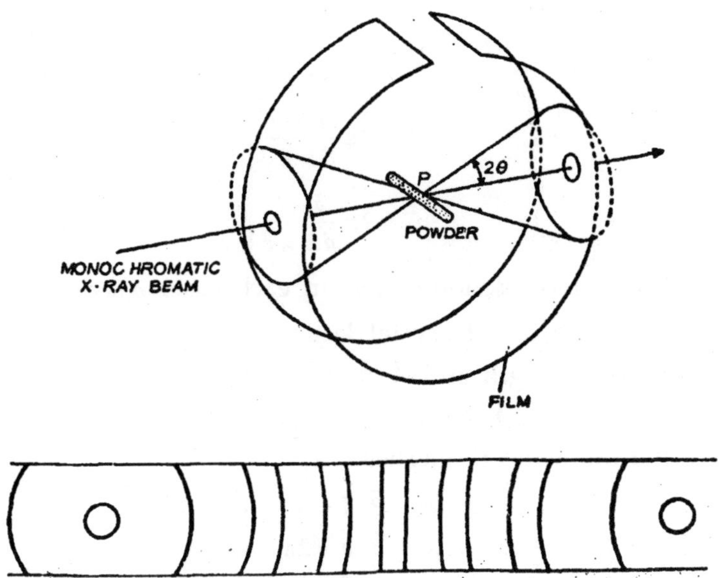

Effect of Reciprocal Lattice

In a crystal lattice there exist many sets of planes, with different orientations and spacings, which can cause diffraction. If we draw, from a common origin, normals to all sets of planes, the length of each normal being proportional to the reciprocal of the interplanar spacing of the corresponding set, then the end-points of the normals form a lattice which is called the 'reciprocal lattice'.

Each point in the reciprocal lattice preserves the characteristics of the set of planes which it represents. Its direction with respect to the origin represents the orientation, of the planes, and its distance from the origin represents the interplanar spacing of the planes.

Reciprocal Lattice Vectors : If $\vec{a}, \vec{b}, \vec{c}$ are the translation vectors of a (direct) crystal lattice, then the translation vectors $\vec{A}, \vec{B}, \vec{C}$ of the reciprocal lattice are defined by

$$\left.\begin{aligned}\vec{A} &= 2\pi\frac{\vec{b}\times\vec{c}}{\vec{a}.\vec{b}\times\vec{c}}\\ \vec{B} &= 2\pi\frac{\vec{c}\times\vec{a}}{\vec{a}.\vec{b}\times\vec{c}}\\ \vec{C} &= 2\pi\frac{\vec{a}\times\vec{b}}{\vec{a}.\vec{b}\times\vec{c}}\end{aligned}\right\} \quad \text{... }(i)$$

The common denominator* in each case, $\vec{a}.\vec{b}\times\vec{c}$, is the volume of the direct crystal lattice. Thus $\vec{a}, \vec{b}, \vec{c}$ have the dimensions of length, and $\vec{A}, \vec{B}, \vec{C}$ those of $[\text{length}]^{-1}$. From the defining eq. (*i*), $\vec{A}, \vec{B}, \vec{C}$ the axes vectors of the reciprocal lattice, have the following properties :

$$\left.\begin{matrix}\vec{A}.\vec{a}=2\pi & \vec{B}.\vec{a}=0, & \vec{C}.\vec{a}=0,\\ \vec{A}.\vec{b}=0, & \vec{B}.\vec{b}=2\pi, & \vec{C}.\vec{b}=0,\\ \vec{A}.\vec{c}=0, & \vec{B}.\vec{c}=0, & \vec{C}.\vec{c}=2\pi.\end{matrix}\right\} \quad \text{... }(ii)$$

Any arbitrary set of vectors $\vec{a}, \vec{b}, \vec{c}$ of a given crystal lattice leads to the same set of reciprocal lattice points

$$\vec{G} = h\vec{A}+k\vec{B}+l\vec{C}, \quad \text{... }(iii)$$

where *h*, *k*, *l* are integers. Any vector $\vec{G}$ of this form is called a 'reciprocal lattice vector'. It connects any two points of the reciprocal lattice.

The (direct), crystal lattice is a lattice in real or ordinary space ; the reciprocal lattice is a lattice in the so-called Fourier space.

The reciprocal lattice vector $\vec{G}$ and the (direct) crystal lattice vector $\vec{T}\left(=u\vec{a}+v\vec{b}+w\vec{c}\right)$ have the following relationship:

$$\begin{aligned}\vec{G}.\vec{T} &= \left(h\vec{A}+k\vec{B}+l\vec{C}\right).\left(u\vec{a}+v\vec{b}+w\vec{c}\right)\\ &= 2\pi\,(hu+kv+lw) \qquad [\text{from eq. }(ii)]\\ &= 2\pi N,\end{aligned}$$

where N is an integer. The quantity $hu + kv + lw$, being the sum of products of integers, is an integer. It follows that

$$e^{i\vec{G}.\vec{T}} = 1$$

because $$e^{i\,2\pi N} = 1$$

Properties of Reciprocal Lattice : Every crystal structure has two lattices associated with it, the (direct) crystal lattice' and the reciprocal lattice. The reciprocal lattice have the following properties with respect to the crystal lattice :

(i) *Every vector of the reciprocal lattice is normal to a set of lattice planes of the crystal lattice.*

Proof. A vector

$$\vec{G}(hkl) = h\vec{A} + k\vec{B} + l\vec{C}$$

will be normal to a plane through the three points $u\vec{a}, v\vec{b}, w\vec{c}$ of the crystal lattice if $\vec{G}$ is normal to any vector lying in that plane. The vectors $u\vec{a} - v\vec{b}, u\vec{a} - w\vec{c}$ and $v\vec{b} - w\vec{c}$ lie in the plane, and so we must have

or $$\left(h\vec{A} + k\vec{B} + l\vec{C}\right).\left(u\vec{a} - v\vec{b}\right) = 2\pi\ (hu - kv) = 0$$

or $$hu = kv.$$

Similarly, we must have

$$hu = lw,$$

and $$kv = lw,$$

These are satisfied, if

$$u = \frac{1}{h},\ v = \frac{1}{k},\ w = \frac{1}{l}.$$

But the Miller indices of the crystal; plane through the points $u\vec{a}, v\vec{b}, w\vec{c}$ are the $(h\ k\ l)$ of the last expression. Hence the vector $\vec{G}$ (*hkl*) of the reciprocal lattice is normal to the plane (*hkl*) of the crystal lattice.

(ii) *The magnitude of every reciprocal lattice vector is inversely proportional to the interplanar spacing of the corresponding set of lattice planes.*

Proof : Let $\vec{r}$ be any vector from the origin of the crystal lattice to a plane (*hkl*). Then the equation of the plane is*

$$\vec{r}\,.\hat{G}(hkl) = d(h\ k\ l)$$

where $\hat{G}$ (*hkl*) is the unit vector normal to the plane (*hkl*).

Let us consider one particular $\vec{r}$, say $\vec{r} = \vec{a}/h$. Then

$$\vec{r}\,.\hat{G}(hkl) = \frac{\vec{a}}{h}.\left(h\vec{A} + k\vec{B} + l\vec{C}\right)$$

$$= \frac{1}{h}(2\pi h) = 2\pi$$

or $$\vec{r}\,.\hat{G}(hkl) = \frac{2\pi}{G(hkl)}. \qquad \left[\because \vec{G} = G\hat{G}\right]$$

Comparing it with the top expression, we get

$$\frac{2\pi}{G(hkl)} = d(hkl)$$

or $$G\,(hkl) \propto \frac{1}{d(hkl)}.$$

(iii) *The volume of a unit cell of the reciprocal lattice is inversely proportional to the volume of a unit cell of the crystal lattice.*

The reciprocal lattice cell is described by $\vec{A}, \vec{B}$ and $\vec{C}$. Its volume would be

$$\vec{A}\,.\vec{B}\times\vec{C} = (2\pi)^3\,\frac{\vec{b}\times\vec{c}}{\vec{a}\,.\vec{b}\times\vec{c}}.\frac{\left(\vec{c}\times\vec{a}\right)\times\left(\vec{a}\times\vec{b}\right)}{\left(\vec{a}\,.\vec{b}\times\vec{c}\right)^2}$$

$$= (2\pi)^3\,\frac{\left(\vec{b}\times\vec{c}\right).\left\{\vec{a}\left(\vec{c}\,.\vec{a}\times\vec{b}\right)\right\}}{\left(\vec{a}\,.\vec{b}\times\vec{c}\right)^3}$$

$$= \quad (2\pi)^3 \frac{\left(\vec{b} \times \vec{c}\right).\left\{\vec{a}\left(\vec{a}\,.\vec{b} \times \vec{c}\right)\right\}}{\left(\vec{a}\,.\vec{b} \times \vec{c}\right)^3}$$

$$= \quad \frac{(2\pi)^3}{\vec{a}\,.\vec{b} \times \vec{c}}$$

where $\vec{a}\,.\vec{b} \times \vec{c}$ is the volume of the unit cell of the crystal lattice.

Importance : A diffraction pattern of a crystal is a map of the reciprocal lattice of the crystal. The Bragg's diffraction law can be expressed in a simple way in terms of the reciprocal lattice vector $\vec{G}$ as

$$2\vec{k}\,.\vec{G} + G^2 \;=\; 0$$

where $\vec{k}$ is the wave vector of the incident wave. This equation is very important in wave propagation and elastic scattering in crystal lattices. A similar result arises in the theory of the electron energy band structure of crystals. Thus the conception of reciprocal lattice is very important in the study of lattice dynamics of solids.

Brillouin Zones : *A Brillouin zone is defined as a Wigner-Seitz cell in the 'reciprocal lattice.* The sets of planes which are the perpendicular bisectors of the reciprocal lattice. Vectors are important in crystals, because a wave whose wave-vector drawn from the origin terminates on any of these planes satisfies Bragg's diffraction condition. These planes divide the Fourier space of the crystal into pieces. The central cell formed in this way in the reciprocal lattice is called the 'first Brillouin zone', and is of special importance in the theory of solids. *The first Brillouin zone is thus the smallest volume entirely enclosed by planes which are the perpendicular bisectors of the reciprocal lattice vectors drawn from the origin.*

The first Brillouin zone of an oblique lattice in two dimensions is constructed in Fig. below. We first draw an adequate number of vectors from O to nearby points in the reciprocal lattice. Next we draw lines perpendicular to these vectors at their midpoints. The *smallest* enclosed area is the first Brillouin zone. Any wave whose wave vector drawn from the origin terminates on the boundary of the zone will be diffracted by the crystal.

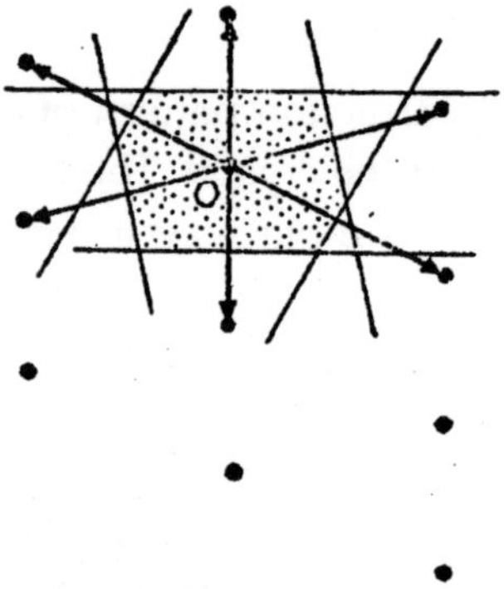

FIRST BRILLOUIN ZONE

Brillouin zones are an essential part of the analysis of the electronic energy-band structure of crystals.

Reciprocal Lattice to Simple Cubic (sc) Lattice : The translation vectors of a simple cubic lattice ($a = b = c$) may be written as

$$\vec{a} = a\hat{i}, \quad \vec{b} = a\hat{j}, \quad \vec{c} = a\hat{k}.$$

The volume of the unit cell is $\vec{a} \cdot \vec{b} \times \vec{c} = a^3$.

The translation vectors of the reciprocal lattice are found from the defining relations :

$$\left[\therefore \hat{j} \times \hat{k} = \hat{i}\right] \quad \left[\therefore \hat{j} \times \hat{k} = \hat{i}\right]$$

$$\left[\therefore \hat{k} \times \hat{i} = \hat{j}\right] \quad \left[\therefore \hat{k} \times \hat{i} = \hat{j}\right]$$

$$\left[\therefore \hat{i} \times \hat{j} = \hat{k}\right] \quad \left[\therefore \hat{i} \times \hat{j} = \hat{k}\right]$$

Thus the reciprocal lattice is itself a simple cubic lattice, now of ' lattice constant $2\pi/a$.

The boundaries of the first Brillouin zone are the planes normal to the six reciprocal lattice vectors $\pm\vec{A}$, $\pm\vec{B}$ and $\pm\vec{C}$ at their mid-points.

$$\left.\begin{aligned} \pm\frac{1}{2}\vec{A} &= \pm\frac{\pi}{a}\hat{i} \\ \pm\frac{1}{2}\vec{B} &= \pm\frac{\pi}{a}\hat{j} \\ \pm\frac{1}{2}\vec{C} &= \pm\frac{\pi}{a}\hat{k}. \end{aligned}\right\}$$

The six planes represented by these equations bound a cube of edge $2\pi/a$ and of volume $(2\pi/a)^3$. This "cube" is the first Brillouin zone of the simple cubic crystal lattice.

PROBLEMS

1. The glancing angle of the (110) plane of a simple cubic crystal (a = 2.814 Å) corresponding to second-order diffraction maximum for the X-rays of wavelength 0.710 Å is calculated as follows.

Solution : The distance between successive lattice planes defined by Miller indices h, k, l in a simple cubic lattice is given by

$$d_{hkl} = \frac{a}{\sqrt{(h^2 + k^2 + l^2)}}.$$

Here $a = 2.814$ Å, $h = 1$, $k = 1$, $l = 0$.

$$\therefore \qquad d_{110} = \frac{2.814\ \text{Å}}{\sqrt{2}} = 1.990\ \text{Å}$$

The Bragg's equation for X-ray diffraction is

$$2\, d_{110} \sin\theta = n\lambda.$$

This gives

$$\sin\theta = \frac{n\lambda}{2d_{110}}$$

Here n (order) = 2, λ = 0.710 Å

and d_{110} = 1.990 Å.

$$\therefore \quad \sin\theta = \frac{2 \times 0.710}{2 \times 1.990} = 0.357$$

or $\theta = \sin^{-1}(0.357) = 21°$

2. *The K_α line from the anticathode of an X-ray tube ... reflected in first order at an angle of 6° by a crystal of NaCl mounted on a Bragg spectrometer. In X. U. the wavelength of K_α radiation is calculated. Density of NaCl = 2170 kg/m³, molecular weight of NaCl = 58.46 and Avogadro number N = 6.02 × 10²⁶/kg-molecule.*

Solution : The interplanar spacing d of the crystal, in terms of molecular weight M and density ρ, is given by

$$d = \left(\frac{M}{2N\rho}\right)^{\frac{1}{3}}$$

Putting the given values :

$$d = \left[\frac{58.46 \text{ kg / kg-molecule}}{2 \times \left(6.02 \times 10^{26} \text{ / kg-molecule}\right) \times 2170\, kg / m^3}\right]^{\frac{1}{3}}$$

$$= [22.37 \times 10^{-30} \text{ m}^3]^{1/3}$$

$$= 2.818 \times 10^{-10}\text{m}.$$

The Bragg's equation for specular reflection is

$$2d \sin\theta = n\lambda.$$

Here θ = 6° and n (order) = 1.

$$\therefore \qquad \lambda = 2d \sin\theta = 2 \times (2.818 \times 10^{-10}\text{ m}) \times \sin 6°$$
$$= 2 \times (2.818 \times 10^{-10}\text{ m}) \times 0.1045$$
$$= 0.5890 \times 10^{-10}\text{m}$$
$$= 589.0 \times 10^{-13}\text{ m}$$
$$= 589.0 \text{ X.U.} \qquad (1 \text{ X.U.} = 10^{-13}\text{ m})$$

3. Electrons are accelerated to 344 volts and are reflected from a crystal. The first reflection maximum occurs when glancing angle is 60°. The interplanar spacing of the crystal is determined. Given: h = 6.62 × 10^{-34} joule-sec, e = l.6 × 10^{-19} coulomb and electron mass m = 9.l × 10^{-31} kg.

Solution : The wavelength λ associated with an electron of mass m moving with (non-relativistic) velocity v is given by

$$\lambda = \frac{h}{mv}.$$

If K be the kinetic energy of the electron, then

$$K = ½\, mv^2 = eV,$$

where V is the accelerating voltage. From this

$$v = \sqrt{\left(\frac{2eV}{m}\right)}$$

and so
$$\lambda = \frac{h}{\sqrt{(2meV)}}$$

Substituting the given values :

$$\lambda = \frac{6.62 \times 10^{-34}\text{ joule-sec}}{\sqrt{\left[2 \times (9.1 \times 10^{-31}\text{kg}) \times (1.6 \times 10^{-19}\text{ coulomb}) \times (344 \text{ volt})\right]}}$$

$$= 0.66 \times 10^{-10}\text{ meter} = 0.66 \text{ Å}.$$

Now, Bragg's equation for reflection is

$$2d \sin\theta = n\lambda.$$

For the 'first' reflection, $n = 1$ and $\theta = 60°$.

$$\therefore \qquad 2d \sin 60° = \lambda = 0.66 \text{ Å}.$$

or
$$d = \frac{0.66 \text{ Å}}{2 \sin 60°}$$

$$= \frac{0.66 \text{ Å}}{2 \times 0.866} = 0.38 \text{ Å}$$

4. The angle of first order diffraction maximum when thermal neutrons (kinetic energy = 0.02 eV) undergo diffraction from a crystal plane (100) is calculated. The Bragg's spacing of the crystal is 2.0 Å.

The mass of neutron is 1.00898 amu and h = 6.62 × 10^{-34} joule-sec. Given : 1 eV = 1.6 × 10^{-19} joule and 1 amu = l.66 × 10^{-27} kg.

Solution : The de Broglie wavelength of neutron, in usual notations, is

$$\lambda = \frac{h}{mv}$$

If K be the kinetic energy, then

$$K = \frac{1}{2} mv^2$$

$$\text{and so } \lambda = \frac{h}{\sqrt{(2mK)}}.$$

Substituting the given values :

$$\lambda = \frac{6.62 \times 10^{-34} \text{ joule-sec}}{\sqrt{\left[2 \times \left(1.00898 \times 1.66 \times 10^{-27} \text{ kg}\right) \times \left(0.02 \times 1.6 \times 10^{-19} \text{ joule}\right)\right]}}$$

$$= 20 \times 10^{-10}\text{m} = 2.0 \text{ Å}.$$

The Bragg's equation for crystal diffraction is

$$2d \sin \theta = n\lambda.$$

Here $d = 2.0$ Å,

n(order) $= 1$

and $\lambda = 2.0$ Å.

$$\therefore \quad \sin \theta = \frac{n\lambda}{2d}$$

$$= \frac{1 \times 2.0 \text{ Å}}{2 \times 2.0 \text{ Å}} = 0.5$$

or $\theta = \sin^{-1}(0.5) = 30°$.

5. A two-dimensional lattice has the basis vectors $\vec{a} = 2\hat{i}$, $\vec{b} = \hat{i} + 2\hat{j}$. The basis vectors of the reciprocal lattice is found as follows.

Solution : Suppose the third vector $\vec{c}$ is parallel to the z-axis. Let $\vec{c} = \hat{k}$.

The vectors of the reciprocal lattice are

$$\vec{A} = 2\pi \frac{\vec{b} \times \vec{c}}{\vec{a} \cdot \vec{b} \times \vec{c}}$$

$$\vec{B} = 2\pi \frac{\vec{c} \times \vec{a}}{\vec{a} \cdot \vec{b} \times \vec{c}}.$$

Now,

$$\vec{b} \times \vec{c} = \left(\hat{i} + 2\hat{j}\right) \times \hat{k}$$

$= \hat{i} \times \hat{k} + 2\hat{j} \times \hat{k}$ (vector product is distributive)

$= -\hat{j} + 2\hat{i}$, (rule of vector product)

$$\vec{c} \times \vec{a} = \hat{k} \times 2\hat{i} = 2\hat{j},$$

$$\text{and} \qquad \vec{a}\,.\,\vec{b}\times\vec{c} = \begin{vmatrix} a_x & a_y & a_z \\ b_x & b_y & b_z \\ c_x & c_y & c_z \end{vmatrix}$$

$$= \begin{vmatrix} 2 & 0 & 0 \\ 1 & 2 & 0 \\ 0 & 0 & 1 \end{vmatrix} = 4.$$

Making these substitutions, we get

$$\vec{A} = \pi\hat{i} - \frac{1}{2}\pi\hat{j}\,;\ \vec{B} = \pi\hat{j}$$

The crystal lattice vectors $\vec{a}, \vec{b}$; and the reciprocal lattice vectors $\vec{A}, \vec{B}$ have been sketched in the Fig. $\vec{A}$ and $\vec{B}$ are perpendicular to sets of planes in the crystal lattice, namely, to the lines parallel to $\vec{b}$ and $\vec{a}$ respectively.

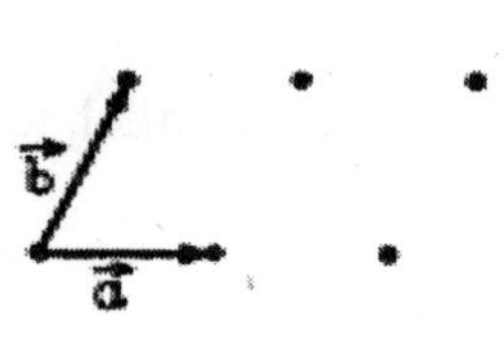

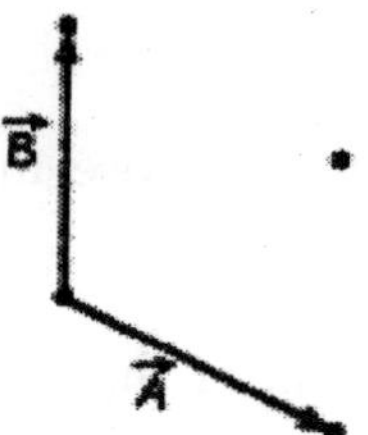

CRYSTAL LATTICE RECIPROCAL LATTICE

12

Amorphous Solids

The Crystalline

Role of Amorphous Solids

A solid consists of atoms, or ions, or molecules packed closely together. In most of the solids the atoms, ions, or molecules are arranged into regular, orderly and periodic three-dimensional patterns. Such solids are called 'crystals'. Thus a crystal is characterised by having a *long-range order* in its structure. This gives a symmetry to the external shape of the crystals.

Solids which lack regularity in the arrangement of their constituent particles are known as 'amorphous solids'. Such solids may, however, show *short-range order* in their structures. Glass, concrete, paper, pitch and plastic are amorphous solids.

Grouping of Crystals

The constituent particles (atoms, ions or molecules) in a crystal are bound together by 'electrostatic' forces, but the distribution of charges in different crystals are qualitatively different. Therefore, the nature of binding varies from crystal to crystal. The difference in bindings is closely connected with the difference in the mechanical, electric and magnetic properties of the crystals.

The binding is characterised by the 'cohesive energy' of the crystal. It is the binding energy per atom or molecule, that is, the energy per atom or molecule that would be liberated in the formation of the crystal from individual neutral atoms or molecules.

The crystals have been classified into five categories according to the binding in them :

1. Ionic Crystals bound by strong electrostatic attraction between ions of opposite kinds.

   Examples: NaCl, CsCl, KBr, etc.

2. Covalent Crystals bound by sharing of electrons between atoms.

   Examples : C (diamond), Ge, Si, etc.

3. Metallic Crystals bound by electrostatic attraction between the lattice of ion cores and the free electron gas.

   Examples : Na, Al, Cu, etc.

4. Inert-gas (or molecular) Crystals bound by weak Van der Waals forces.

   Examples : Solid A, solid CH_4, etc.

5. Hydrogen-bonded Crystals bound by hydrogen bonds.

   Examples : H_2O (ice), HF, etc.

Use of Ionic Crystals

In ionic crystals, electrons are transferred from one type of atoms (which lose electrons readily) to the other type (which have high affinity for electrons). Thus the atoms become positive and negative ions. These ions arrange themselves in a configuration in which Coulomb attraction between ions of opposite signs is stronger than the Coulomb repulsion between ions of the same sign.

Thus the "ionic bond" results from the electrostatic interaction of oppositely charged ions. The configuration remains stable in spite of the net electrostatic attraction between the ions because electron shells of different ions cannot mesh together due to the forces present by the action of the exclusion principle.

Examples : The examples of ionic crystals are sodium chloride (NaCl), cesium chloride (CsCl), lithium fluoride (LiF), potassium bromide (KBr), etc.

In the NaCl crystal the outermost (3*s*) electron of the sodium atom ($1s^2 2s^2 2p^6 3s$) is transferred to the 3*p* subshell of the chlorine atom ($1s^2 2s^2 2p^6 3s^2 3p^5$). The Na^+ and Cl^- ions so formed are arranged in a pattern shown in Fig. below. The ions of either kind are located at the corners and at the centres of the faces of an assembly of cubes. Such a crystal structure is called 'face-centered cubic'.

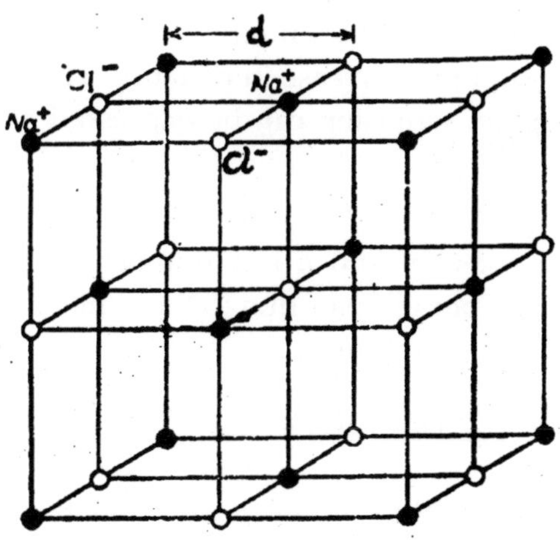

Energy of Cohesion

The cohesive energy of an ionic crystal is the energy that would be liberated in the formation of the crystal from individual neutral atoms, or the energy needed to break the crystal into individual atoms. Let us calculate it:

The Coulomb energy of attraction between two ions separated by a distance r is

$$U_{att} = -\frac{e^2}{4\pi\epsilon_0 r}.$$

In the crystal each ion interacts with all its neighbouring ions of same as well as of opposite signs. To take this into account, the net energy of attraction per ion-pair may be written as

$$U_{att} = -\alpha\frac{e^2}{4\pi\epsilon_0 r},$$

where α is called the 'Madelung constant' of the crystal. For crystal structures, α lies between 1.6 and 1.8.

As mentioned above, two ions cannot continuously approach each other under Coulomb attraction on account of the exclusion principle. When they are at a certain small distance apart, they begin to repel each other with a force which rises very rapidly with the decrease in distance. The energy of this (non-electrostatic) repulsion can be expressed as

$$U_{rep} = \frac{B}{r^n}$$

where B is a constant and n is a number ($n \approx 9$). The total potential energy per ion-pair due to their interactions with all the other ions is therefore

$$U = U_{att} + U_{rep} = -\alpha\frac{e^2}{4\pi\epsilon_0 r} + \frac{B}{r^n}.$$

At the equilibrium separation r_0 of the ions U must be a minimum.

This means that at $r = r_0$, $dU/dr = 0$.

Hence

$$\left(\frac{dU}{dr}\right)_{r = r_0} = \frac{\alpha e^2}{4\pi\epsilon_0 r_0^2} - \frac{nB}{r_0^{n+1}} = 0$$

or $$B = \frac{\alpha e^2}{4\pi\epsilon_0 n} r_0^{n-1}.$$

The potential energy at $r = r_0$ is thus

$$U = -\alpha \frac{e^2}{4\pi\epsilon_0 r_0} + \frac{\alpha e^2}{4\pi\epsilon_0 n r_0}$$

or $$U = -\frac{\alpha e^2}{4\pi\epsilon_0 r_0}\left(1 - \frac{1}{n}\right).$$

This is the total potential energy per ion-pair in the crystal. In NaCl crystal, the equilibrium distance between ions is $r_0 = 2.81$Å, and $\alpha = 1.748$. Taking $n = 9$, the potential energy per ion-pair is

$$U = -\frac{\alpha e^2}{4\pi\epsilon_0 r_0}\left(1 - \frac{1}{n}\right)$$

$$= -\frac{(9 \times 10^9 \text{ newton-m}^2/\text{coul}^2) \times 1.748 \times (1.6 \times 10^{-19} \text{coul}^2)}{(2.81 \times 10^{-10} \text{ m})}\left(1 - \frac{1}{9}\right)$$

$$= 1.27 \times 10^{-18} \text{ joule}$$

$$= -7.97 \text{ eV}.$$

Hence the contribution *per ion* to the cohesive energy of the crystal is half of this, *i.e.*, – 3.99 eV. This is called the 'lattice energy per ion' in the NaCl crystal.

In fact, some energy is needed to transfer an electron from a Na atom to a Cl atom to form $Na^+ - Cl^-$ ion pair. This is equal to the difference between the + 5.14 eV ionisation energy of Na and the – 3.61 eV electron affinity of Cl, *i.e.*, +1.53 eV. Each atom thus contributes + 0.77 eV to the cohesive energy. The total cohesive energy per atom in the NaCl crystal is therefore

$$U_{cotesive} = -3.99 + 0.77 = -3.22 \text{ eV/atom.}$$

This is greater than the binding energy per atom (2.1 eV/atom) in an actual NaCl molecule.

Properties : An ionic bond is a strong bond because the binding forces are of electrostatic origin. In view of the cohesive energies of ionic solids, which are of the order of several electron-volts, we expect the following properties of these solids:

(i) Most ionic solids are hard cubic crystals, owing to the strength of the bonds between their constituent ions.

(ii) They are usually brittle, since the slipping of atoms past one another (which accounts for the ductility of metals) is prevented by the ordering of positive and negative ions.

(iii) They have high melting and vaporizing points because of their cohesive energies of the order of several electron volts. In order to vaporise an ionic solid we must heat it to a temperature T such that $kT \approx 1$ eV. Obviously, T must be quite high.

(iv) They are transparent to visible radiation. A photon of visible light will interact with a material only which it can excite an electron from its ground state to an excited state. Since the ionic solids all have filled shells, an electron must be excited *from one shell to the next*. The energy required for this is much larger the visible-photon energy. So the visible photons pass right through ionic solids.

(v) They absorb strongly in the infrared. This is because an ion in a solid is capable of oscillation with infrared frequencies"

(vi) They are poor conductors of electricity, because they do not have free electrons. However, in contrast to metals, the conductivity of ionic solids increases with rise in temperature.

(vii) Ionic solids are usually soluble in polar liquids like water. The water molecule has an electric dipole moment, which exerts an attractive force on the charged ions, breaking the ionic bonds and dissolving the solid.

(viii) Ionic solids do not distinguish one direction in space from another. This is because the *filled* shells of the ions are spherically symmetric and so the ionic bonds are *not* directional.

Crystals of Co-valency

In covalent crystals, one or more electrons are detached from two adjacent atoms and are *shared* equally by both of them. The simplest example of electron sharing is the H_2 molecule in which two electrons are shared by the two atoms (Fig.). As these electrons circulate, they spend more time between the atoms (in fact between the protons) than elsewhere and this produces an attractive force.

The two electrons are said to form a 'covalent' bond. The spin of the electrons in the bond are antiparallel, *i.e.*, the electrons form a pair.

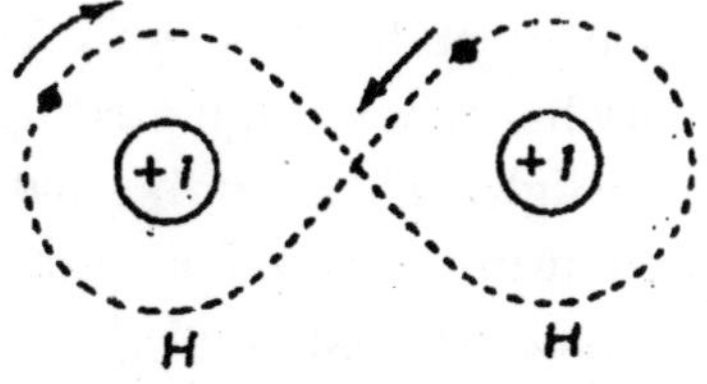

Examples : Diamond (C), silicon (*Si*), germanium (*Ge*), gray tin (*Sn*), silicon carbide (*SiC*) and zinc sulphide (ZnS) are the examples of covalent crystals.

The below given figure shows how the solid structure of diamond is constructed. A central carbon atom is bound to four close neighbours at tetrahedral angles, with each of which it shares electrons in covalent bonds. This happens with each carbon atom in the extended structure. Silicon and germanium also have a diamond like structure. In SiC each atom is surrounded by four atoms of the other kind in the same tetrahedral structure as diamond. (Fig.).

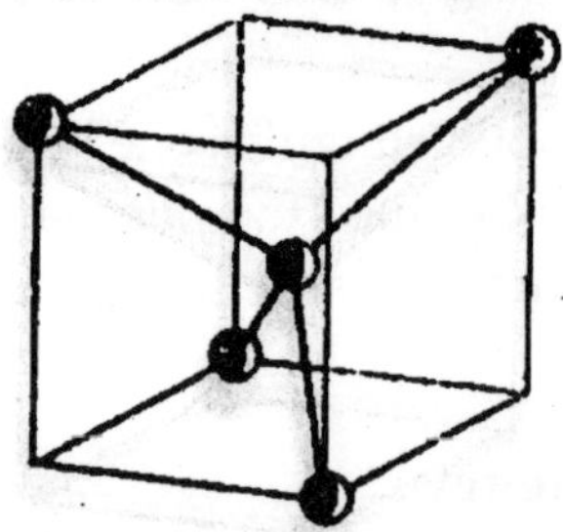

Properties : A covalent bond is a strong bond. Cohesive energies of 6 to 12 eV/atom are typical of covalent crystals. As an example, the bond between two carbon atoms in diamond has a cohesive energy of 7.3 eV with respect to separated neutral atoms. This is comparable with the cohesive energies (bond strengths) in many ionic crystals. As such there is no sharp distinction between ionic and covalent crystals. In fact there is a continuous range of crystals between the ionic and covalent limits.

The actual properties of a given covalent crystal depends on its bond energy. In general, covalent crystals with large bond energies are very hard, have high melting points, and are transparent to visible light. Those with small bond energies may have very different properties. For example, germanium and tin are soft (compared to diamond), melt at much lower temperatures and are reflective to visible light.

The conductivity of covalent crystals varies over a wide range, and increases with increasing temperature.

In contrast to ionic crystals, the covalent crystals have strong *directional* properties.

Crystals of Metal

Metallic crystals, commonly known as metals, are made up of atoms whose valence electrons are weakly coupled to the nucleus: In the solid state, the atoms are so close that they give up their valence electrons which wander throughout the interior of the metal. Thus *a metal consists of Positive ion cores immersed in a gas of free electrons.* The (metallic) binding results from the attraction between the positive metal ions and the electron gas, which exceeds the mutual repulsion of the electrons in the gas and of the positive ions.

Examples : Most of the elements in the first four groups of the periodic table and the transition elements are metals. Li, Na, Cu, Ag, Zn, Fe are some examples.

Some elements are on the borderline between metallio and covalent crystals. Tin is an example. Above 13.2° C the metal "white tin" exists, while below 13.2° C the covalent solid "gray tin" exists.

Metallic crystals exist in bcc, fcc and hcp structures.

Properties : The characteristic properties of metals are their high electrical and thermal conductivities, surface luster, ductility and opacity.

(i) The high conductivity follows from the ability of the free electrons to move throughout the metal, (In contrast, the electrons in ionic and covalent crystals are bound to particular atoms or pairs of atoms).

(ii) When light shines on a metal, the free electrons oscillate under the electromagnetic field of the incident light and become sources of light. This gives the metal its surface luster.

(iii) Since the metallic bonds are not localised between adjacent atoms, the atoms of a metal can be re-arranged in position without rupturing the crystal. This *explains* ductility of metals

(iv) Metallic bonds are weaker than ionic and covalent bonds. For example, cohesive energy in sodium is 1.13 eV/atom and that in zinc 1.35 eV/atom. As a result, metals interact strongly with photons of visible light, and so metals are not transparent. In fact, they are highly reflective.

(v) Because of the lower cohesive energies, metals have relatively lower melting points. Transition metals, however, fall in the upper limit of cohesive energy and so they have higher melting points.

(vi) Since metallic bonds do not depend on any sharing or exchange of electrons between specific atoms, the exact nature of the atoms of the metal is not as important as it is in case of ionic or covalent solids. Hence we can make various metallic alloys by mixing different metals in varying proportions.

Molecular or Inert-gas Crystals : Many solids, or crystals, are composed of *neutral* atoms or molecules, without any transferring, sharing or free movement of electrons which could form ionic, covalent or metallic bonds. These solids remain bound by *much weaker, short-range* attractive forces, known as "Van der Waals forces". These forces arise as a result of fluctuations in the charge distributions of nearby molecules.

Inert-gas solids like solid argon and many organic symmetric-molecule solids like solid methane (CH_4) are the examples of molecular solids.

To investigate the origin of Van der Waals forces, let us consider two identical inert-gas atoms at a separation *r*. The electronic charge distribution in each atom is spherically symmetric *on the average**. However, the electrons are in constant motion around the nucleus. Therefore, *at any instant* one part or other of the atom may have an excess of them, that is, the atom may have an *instantaneous* electric dipole moment. As shown in below given figure for two instants, an instantaneous dipole moment of magnitude p_1 on one atom produces an electric field E of magnitude $2p_1/r^3$ at the centre of the second atom. This field will induce an instantaneous dipole moment $p_2 = \alpha E = 2\alpha p_1/r^3$ on the second atom, where α is the polarizability of the potential energy of the dipole moments is

$$U(r) = -\frac{2p_1p_2}{r^3} = -\frac{4\alpha p_1^2}{r^6}$$

or

$$U(r) = -\frac{C}{r^6}$$

where C is a constant. The mutual energy of the two atoms arising from their (Van der Waals) interaction is thus *negative,* signifying that the force between them is *attractive.* The force itself is equal to —dU/dr and so is proportional to r^{-7}, which means that it drops rapidly with increasing separation. It is this force which bonds the crystals of inert gases and also the crystals of many organic molecules. Ordinary gases like O_2, N_2, H_2 form molecular solids in the solid state.

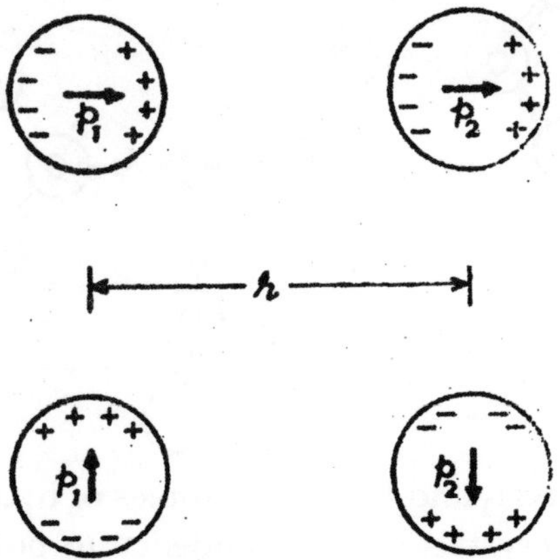

Properties : Van der Waals forces are much weaker than those found in ionic and covalent bonds. This means that little energy is required to break the bonds. Hence the molecular crystals have low melting and boiling points and little mechanical strength. It is easy to deform and compress them. The absence of free electrons makes them very poor conductors of heat and electricity.

Cohesive energies are low, only 0.08 eV/atom in solid argon (melting point —189°C) and 0.1 eV/molecule in solid methane (melting point — 183°C).

Crystals with Hydrogen Bond

In certain crystals an atom of hydrogen attracts two extremely electronegative atoms, particularly F, O and N, thus forming a "hydrogen bond" between them. In the extreme case, the hydrogen atom loses its electron to another atom in the molecular the bare proton forms the hydrogen bond. These bonds are stronger than Van der Waals bonds, but weaker than ionic and covalent bonds.

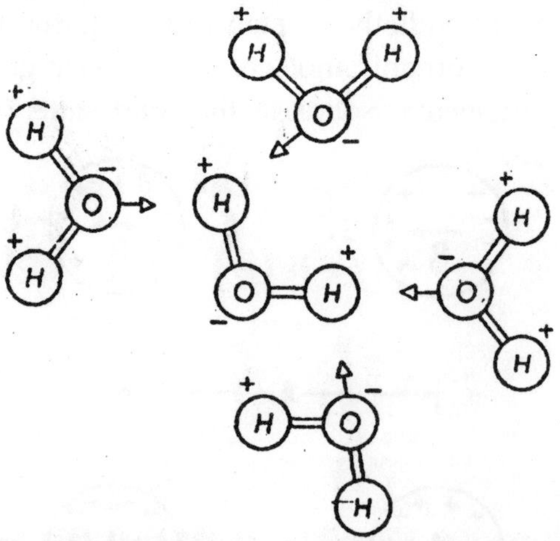

H_2O (ice), NH_3 and HF are examples of hydrogen-bonded crystals. Such crystals show a tendency to polymerise.

In H_2O molecule, the oxygen atom attracts all the electrons of the molecule and becomes the negative end. The two 'fibre' protons form the positive ends, and each attracts the negative oxygen of two adjacent molecules (Fig.). The hexagonal pattern of an ice crystal arises from the tetrahedral arrangement of the four hydrogen bonds each H_2O molecule can participate in.

The following crystals are classified according to their bondings

Na, NaCl, diamond, KBr, Cu, Zn, Si.

Ionic : NaCl, KBr.

Covalent : Diamond, Si.

Metallic ; Na, Cu, Zn.

PROBLEM

1. The ionic cohesive energy and the atomic cohesive energy of CsCl are found out as here. The Madelung constant for CsCl is 1.763 and the separation between adjacent ions is 3.56 Å. The non-electrostatic repulsive energy varies as r^{-n}, where $n = 10.5$. The ionisation energy of Cs is 3.89 eV and the electron affinity of Cl is – 3.61 eV.

Solution: The ionic cohesive energy of CsCl is given by

$$U = -\frac{\alpha e^2}{4\pi\epsilon_0 r_0}\left(1 - \frac{1}{n}\right),$$

where $\frac{1}{4\pi\epsilon_0} = 9 \times 10^9$ N-m^2/C^2.

Substituting the given data for CsCl,

$$U = -\frac{\left(9 \times 10^9 \; N - \text{m}^2/\text{C}^2\right)(1.763)\left(1.6 \times 10^{-19} \; \text{C}\right)^2}{3.56 \times 10^{-10} \; \text{m}}\left(1 - \frac{1}{10.5}\right)$$

$$= -10.3 \times 10^{-19} \text{ J}$$

$$= -\frac{10.3 \times 10^{-19} \text{ J}}{1.6 \times 10^{-19} \text{ j/eV}}$$

$$= -6.45 \text{ eV}.$$

This represents the energy needed to remove Cs^+ and Cl^- ions from the crystal. To form Cs^+ from Cs atom requires 3.89 eV, and to form Cl^- from neutral Cl atom releases 3.61 eV. Thus, given Cs^+ and Cl^-, we supply 3.61 eV to form Cl^-, and get back 3.89 eV when we form Cs. The energy needed to form the neutral atoms from the crystal is

$$-6.45 \text{ eV} - 3.61 \text{ eV} + 3.89 \text{ eV} = -6.17 \text{ eV}.$$

This is the atomic cohesive energy.

13

Monatomic Vibrations

Monatomic Linear Crystal

Vibrations of Monatomic Linear Lattices : Vibrations

A monatomic linear crystal lattice is an array of identical, equidistant atoms (or ions), each of mass M (say). When an elastic wave propagates through the crystal, the atoms are set in vibration either parallel or perpendicular to the direction of the wave vector. We shall consider longitudinal elastic vibrations of the atoms. We assume that interaction exists only between nearest neighbours and that Hooke's law is obeyed.

Below given figure shows a linear chain of identical atoms. Let it be infinitely long. The open circles represent the equilibrium positions and the black dots the displaced atoms. Let a be the equilibrium interatomic spacing, and x_0, x_1, x_2, .. x_{n-1} . x_n, x_{n+1}, ... the displacements from the equilibrium positions. According to Hooke's law, the force on an atom due to the displacement of its (nearest) neighbour is proportional to the difference of their displacements. So the net force on the nth atom is

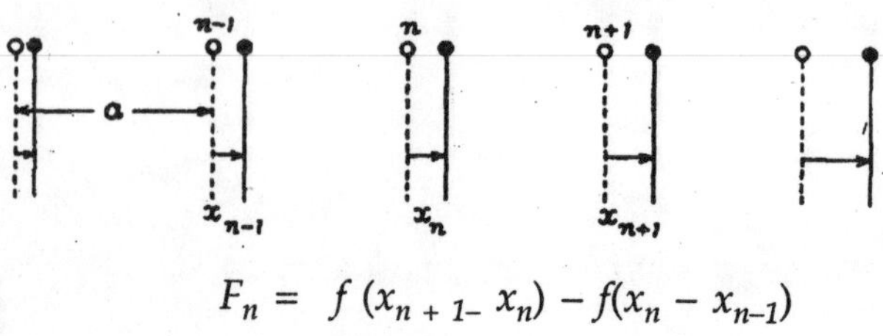

$$F_n = f(x_{n+1-} x_n) - f(x_n - x_{n-1})$$
$$= f(x_{n+1} + x_{n-1} - 2x_n)$$

where f is the force constant for the nearest-neighbour interaction. The equation of motion of the nth atom is then

$$M x_n = f(x_{n+1} + x_{n-1} - 2x_n) \quad ...(i)$$

Let a solution to this equation be of the form of a travelling waver

$$x_n = e^{-iw(t - na/v)},$$

where na is the equilibrium position (x-coordinate) of the nth particle relative to the origin, and v the velocity of wave propagation.

Let us introduce the wave vector K, where

$$K = \frac{2\pi}{\lambda} = \frac{\omega}{v}. \quad [\therefore \omega = 2\pi v \text{ and } v = v\lambda]$$

Then we have

$$x_n = e^{-i(wt - Kna)}. \quad ... (ii)$$

Accordingly,

$$x_{n+1} = e^{-i\{wt - K(n+1)a\}} = e^{iKa} x_n$$

and $$x_{n-1} = e^{-i\{wt - K(n-1)a\}} = e^{-iKa} x_n.$$

Differentiating eq. (ii) twice with respect to t, we get

$$\ddot{x}_n = \frac{d^2 x_n}{dt^2} = (-iw)2\, e^{-i(wt - Kna)} = -w^2 x_n.$$

Substituting for $\ddot{x}_n$, x_{n+1}, x_{n-1} and x_n in eq. (i), and dividing by x_n, we get

$$-M\omega^2 = f\,(e^{iKa} + e^{-iKa} - 2)$$
$$= f\,(e^{iKa/2} - e^{-iKa/2})^2.$$

Now, $\sin \dfrac{Ka}{2} = \dfrac{e^{iKa/2} - e^{-iKa/2}}{2i},$

and so, $\sin^2 \dfrac{Ka}{2} = -\dfrac{1}{4}\,(e^{iKa/2} - e^{-iKa/2})^2.$

$$\therefore \qquad -M\omega^2 = -4f \sin^2 \frac{Ka}{2}$$

or
$$\omega^2 = \frac{4f}{M} \sin^2 \frac{Ka}{2}$$

or
$$\omega = 2\sqrt{\frac{f}{M}}\left|\sin \frac{Ka}{2}\right|. \qquad \text{... }(iii)$$

We take the sign of the square-root so that the angular frequency ω is always positive.

We have thus obtained an expression for the frequency ω of the elastic waves in a monatomic linear lattice in terms of K (that is, in terms of A). The ω-K relationship is shown in Fig. below. It shows that frequency is a periodic function of K, with period $2n/a$. The dashed lines correspond to a continuous string for which $\omega \propto K$.

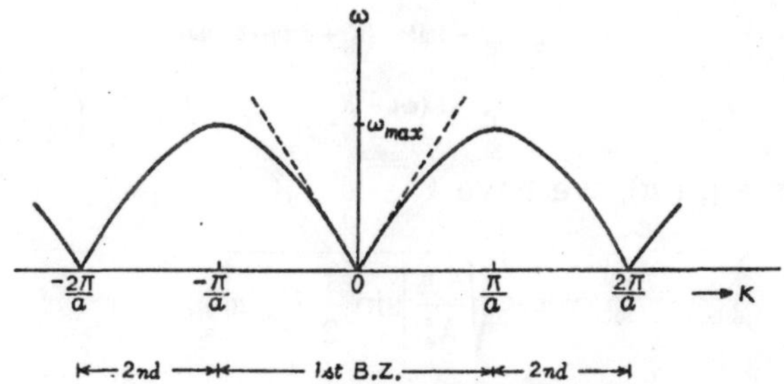

Results : We draw the following results from the above discussion:

(1) For values of K much smaller than interatomic spacing a, that is, for wavelengths large compared with a, we have

$$\sin\frac{Ka}{2} \simeq \frac{Ka}{2} \qquad [\therefore\ Ka < 1]$$

$$\therefore \qquad \omega = 2\sqrt{\frac{f}{M}}\left(\frac{Ka}{2}\right) = \alpha\sqrt{\frac{f}{M}}(K).$$

The result that the frequency is directly proportional to the wave vector is a statement that the velocity of propagation of waves is independent of wavelength. That is, dispersion effects are negligible. This is what happens in a continuous string. Thus, *for long wavelengths,* the periodicity of the lattice is of little importance.

As K increases (wavelength decreases), however, the dispersion effects become important and ω is no longer directly proportional to K.

(2) Another important result is obtained by comparing the solution (it) with another in which K is replaced by K_m where

$$Km = K + \frac{2\pi m}{a} \quad \text{with } m = \pm 1, \pm 2, \dots \dots$$

We have for the displacement of the nth atom

$$x_n = e^{-\iota(\omega t - K_m na)}$$

$$= e^{-\{\omega t - (K + 2\pi m/a)\, na\}}$$

$$= e^{-i(\omega t - Kna)}, \qquad [\therefore\ e^{2\pi i m} = 1]$$

and from eq. (*iii*), we have

$$\omega = \sqrt{\frac{4f}{M}\left|\sin\frac{1}{2}K_m\, a\right|}$$

$$= \sqrt{\frac{4f}{M}\left|\sin\frac{1}{2}a\left(K + \frac{2\pi m}{a}\right)\right|}$$

$$= \sqrt{\frac{4f}{M}} \left| \sin \frac{1}{2} Ka \right| .$$

Thus the solutions (*ii*) and the frequencies (*iii*) corresponding to K and K_m are identical. It means that the state of vibration of the lattice corresponding to a wave vector K is the same as that for any of the wave vectors $K + 2\pi m/a$. Therefore, to find a unique relation between the state of vibration and the wave vector, the wave vector K must be restricted to a range of values $2\pi/a$. We specify the range of independent values of K by

$$-\frac{\pi}{a} \leq K \leq \frac{\pi}{a}.$$

We have taken both positive and negative values of K because waves can propagate to the right or to the left. This range of K values is referred to as the "first Brillouin zone" of the linear lattice. The extreme values of K in this zone are

$$K_{max} = \pm \frac{\pi}{a},$$

where K_{max} may be of the order of 108 cm^{-1}. (In an elastic continuum $a \to 0$ and $K_{max} \to \pm \infty$).

At the boundaries $K_{max} = \pm \frac{\pi}{a}$ of the Brillouin zone the solution $x_n = e^{-i(\omega t - Kna)}$ becomes

$$x_n = e^{-i\omega t} e^{\pm in\pi} = e^{-i\omega t} (-1)n$$

which represents a "standing" wave (rather than a travelling wave) in which alternate atoms vibrate in opposite phases. This situation is equivalent to Bragg reflection of X-rays. When the Bragg condition is satisfied, a wave cannot propagate in a lattice, but through successive reflections back and forth , a standing wave is set up. Here the critical value $K_{max} = \pm \frac{\pi}{a}$ satisfies the Bragg condition $2d \sin \theta = n\lambda$. Here we have

$\theta = \frac{\pi}{2}$ (linear lattice), $d = a, n = 1$, so that

$$\lambda = \lambda_{min} = 2a$$

which corresponds to $K_{max} = \pm \pi/a$.

3. The frequency corresponding to zone boundaries ($K = K_{max} = \pm \pi/a$) is, according to eq. (*iii*),

$$\omega_{max} = 2\sqrt{\frac{f}{M}}$$

or $$V_{max} = \frac{1}{\pi}\sqrt{\frac{f}{M}}. \qquad [\therefore \omega = 2\pi v]$$

Thus there exists a maximum frequency v_{max} which can be propagated through the chain. Hence a monatomic linear lattice acts as a low-pass filter which transmits only in the frequency range between zero and v_{max}.

$$v_{max} = \frac{v}{\lambda_{min}} = \frac{v}{2a}.$$

Now a $\simeq 10^{-8}$ cm and the velocity of sound in solids is $\sim 10^5 - 10^6$ cm/sec, so that

$$v_{max} \simeq 10^{13}\ \text{sec}^{-1}.$$

Hence the cut-off frequency lies in the infrared region.

In contrast with this, the continuous string has no frequency limit.

Vibrational Modes of a Finite Monatomic Linear Lattice : Let us consider an array *of* $(N + 1)$ identical atoms , numbered from 0 to N. Suppose that the two end atoms are fixed, so that $(N - 1)$ atoms are mobile. Let a be the equilibrium interatomic spacing , and $x_0, x_1, x_2,......x_N$ the instantaneous displacements from the equilibrium positions. Assuming Hooke's law to be obeyed and considering only nearest-neighbour interaction, the equation of motion of the nth atom is

$$M\ddot{x}_n = f(x_{n+1} - x_n) - f(x_n - x_{n-1})$$

$$= f(x_{n+1} + x_{n-1} - 2x_n) \quad \text{... } (i)$$

where M is the mass of an atom and f is the force constant for the nearest-neighbour interaction.

The general solution of eq. (*i*) can be written as a sum of two travelling waves, one propagating to the right and the other to the left :

$$x_n = [A_1\, e^{i(Kna + \beta_1)} + A_2\, e^{\,i(-Kna + \beta_2)}]\, e^{-i\omega t}, \quad \text{... } (ii)$$

where A_1 and A_2 are amplitudes, β_1 and β_2 are phase angles and K is wave vector ($K = 2\pi/\lambda$). The boundary conditions are

$$x_0 = 0$$

$$x_N = 0 \text{ for all } t.$$

The first condition, when substituted in (it), requires

$$A_1 = -A_2$$

and $\beta_1 = \beta_2$.

Since the phase angles are equal, we can take $B_1 = B_2 = 0$. The remaining solution in (*ii*) is

$$x_n = A_1\,(e^{iKna} - e^{-ikna})\, e^{-i\omega t}.$$

Taking the real part only, we have

$$x_n = 2A_1 \sin Kna \sin \omega t. \quad \text{... } (iii)$$

This represents a "standing" wave. It leads to the same dispersion relation (ω-K relationship)

$$\omega^2 = \frac{4f}{M}\sin^2\frac{Ka}{2}$$

as the travelling wave solution, with the difference that K is limited to positive values ranging from 0 to π/a.

The second boundary condition imposed on (*iii*) gives

$$0 = 2A_1 \sin KNa \sin \omega t$$

or $\sin KNa = 0$

or $$K = \frac{\pi}{Na}\, j, \qquad \text{... } (iv)$$

where j is an integer. However, $j = 0$ must be excluded because it corresponds to $K = 0$ which means that all atoms are at rest.

The maximum value of K is $\frac{\pi}{a}$ which gives $j_{max} = N$. This value must also be excluded because it yields in (*iii*) $x_n = 0$ for all t, that is, again all atoms are at rest. We therefore conclude that

$$j = 1, 2, 3, \text{.........}, (N - 1). \qquad \text{... } (v)$$

Thus the boundary conditions select a discrete set of K values given by (*iv*) and (*v*). We see that, *there are just as many modes of vibration (K-values) as there are mobile atoms.*

To each value of K there corresponds a frequency ω_K. Hence the frequency spectrum consists of $(N - 1)$ discrete lines.

Vibrational Modes of a Diatomic Linear Lattice: For crystals with more than one atom per primitive cell, the vibrational spectrum shows new features. For each polarisation mode in a given direction of propagation, the dispersion relation $\omega - K$ gives rise to two branches, known as 'acoustical branch and optical branch'.

Let us consider a linear diatomic lattice, with atoms of two kinds arranged alternately (Fig.). Let the atoms of mass M_1 be even-numbered, and those of mass M_2, odd-numbered, and Let a be the distance between nearest neighbours, so that $2a$ is the repeat distance.

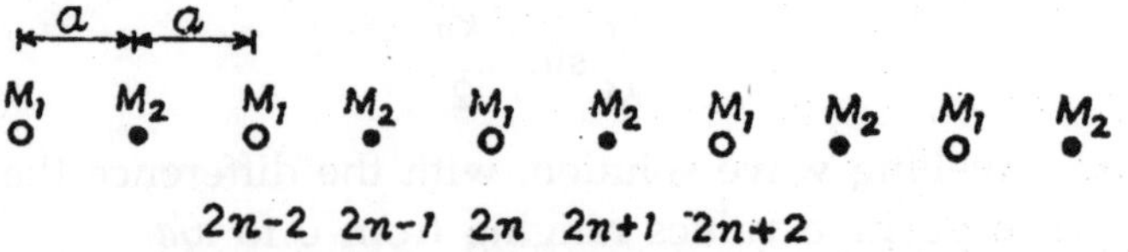

a a

M_1 M_2 M_1 M_2 M_1 M_2 M_1 M_2 M_1 M_2

$2n-2$ $2n-1$ $2n$ $2n+1$ $2n+2$

We assume that each atom interacts only with its nearest neighbour and that the force constants are identical between all nearest-neighbour pairs. The equations of motion are

$$M_1 \ddot{x}_{2n} = f(x_{2n+1} + x_{2n-1} - 2x_{2n})$$

and $$M_2 \ddot{x}_{2n+1} = f(x_{2n+2} + x_{2n} - 2x_{2n+1}). \qquad ... (i)$$

where f is the force constant and x_{2n} is the displacement of the 2nth particle from the equilibrium position.

Let the solutions to these equations be of the form of a travelling wave :

$$x_{2n} = A\, e^{-i(\omega t - 2nKa)}$$

and $$x_{2n+1} = B\, e^{-i\{\omega t - (2n+1)Ka\}} \qquad ... (ii)$$

where K is the wave vector of a particular mode of vibration, and A and B are the amplitudes corresponding to the atoms M_1 and M_1 respectively. Substituting (*ii*) in (*i*), we get

$$(M_1 \omega^2 - 2f)A + 2\,Bf\cos Ka = 0\,)$$

and $$(M_2\, \omega^2 - 2f)\,B + 2Af\cos Ka = 0.... \; (iii)$$

These equations have non-vanishing solutions for A and B only if the determinant of the coefficients of A and B vanishes, that is,

$$\begin{vmatrix} (M_1\,\omega^2 - 2f) & 2f\cos Ka \\ 2f\cos Ka & (M_2\,\omega^2 - 2f) \end{vmatrix} = 0$$

or $$(M_1\omega^2 - 2f)\,(M_2\omega^2 - 2f) - 4f^2\cos^2 Ka = 0$$

or $$M_1 M_2\,\omega^4 - 2f\omega^2)\,(M_1 + M_2) + 4f^2\,(1 - \cos^2 Ka = 0$$

giving finally

$$\omega^2 = f\left(\frac{1}{M_1} + \frac{1}{M_2}\right) \pm f\sqrt{\left[\left(\frac{1}{M_1} + \frac{1}{2}\right)^2 - \frac{4\sin^2 Ka}{M_1 M_2}\right]} \qquad ... (iv)$$

This is the dispersion relation for the linear diatomic lattice. The form of the relation shows that, as in the monatomic case, the frequency ω is a periodic function of the wave vector K.

Since ω should be positive, each value of ω^2 leads to a single value for *at*. Thus, in contrast to the monatomic lattice, there are now two angular frequencies ω_+ and ω_- corresponding to a single value of *K*. In a ω – *K* plot (Fig.) this leads to two "branches", the one corresponding to ω_+ is called the 'optical' branch and that corresponding to ω_- is called the 'acoustical' branch.

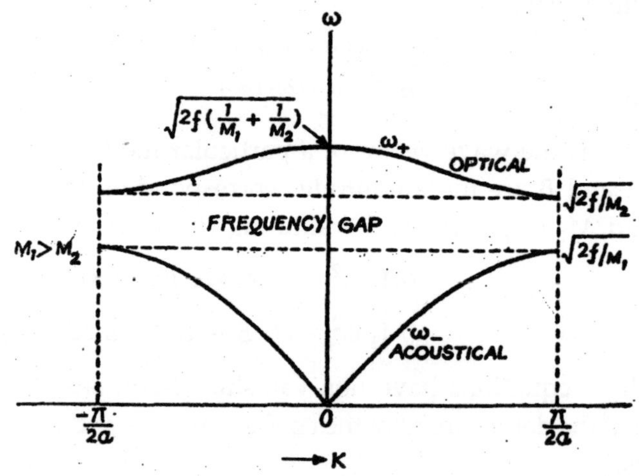

Let us discuss the two branches. For *K* = 0, two roots of (*iv*) are

$$\omega + = \sqrt{\left[2f\left(\frac{1}{M_1}+\frac{1}{M_2}\right)\right]} \quad \text{... }(v)$$

and $\quad \omega - = 0.$

The range of the first Brillouin zone is

$$\frac{\pi}{2a} \leq K \leq \frac{\pi}{2a},$$

where *la* is the repeat distance of the lattice. At $K = K_{max} = \pm \frac{\pi}{2a}$, the angular frequencies are

$$\omega + = \sqrt{\left(\frac{2f}{M_1}\right)}, \ \omega- = \sqrt{\left(\frac{2f}{M_2}\right)}.$$

The larger the mass ratio M_1/M_2, the wider the 'frequency gap' between the two branches. The existence of a frequency (or energy) gap is a characteristic feature of elastic waves in diatomic lattices.

Let us now investigate the physical difference between the optical and acoustical branches. The particle displacements for the optical and acoustical modes of vibration of (transverse) waves in diatomic linear lattice are shown in Fig. below.

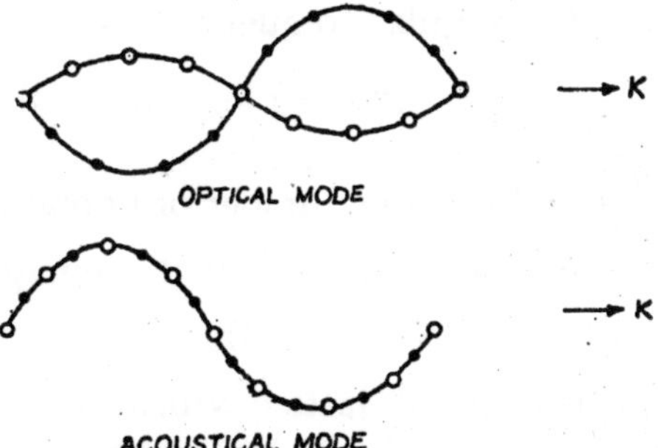

Let us calculate the ratio of the amplitudes A and B in the two cases. Let us see the situation for $K = 0$.

From eq. (*iii*) and (*v*), for $K = 0$, it follows that

$$\frac{A}{B} = -\frac{M_2}{M_1} \qquad \text{(for optical branch)}$$

and $$A = B. \qquad \text{(for acoustical branch)}$$

In the optical mode, the two types of atoms vibrate against each other with amplitudes in inverse ratio of their masses so that their centre of mass remains fixed. If the two atoms carry opposite charges, as in ionic crystals, then a vibration of this type can be excited by the electric field of a light wave. Hence the name optical branch.

In the acoustical mode, the two types of atoms (and their centre of mass) move together as happens in long-wavelength acoustical waves. Hence the name acoustical branch.

Quantisation of Lattice Vibrations—Phonons : The energy of a lattice vibration is quantised. The quantum of energy is called a "phonon" (in analogy with the photon of the electromagnetic wave). Thus, elastic waves in crystals are made up of phonons. Thermal vibrations in crystals are thermally-excited phonons (like the thermally-excited photons of black-body electromagnetic radiation).

The energy of each mode of vibration in an elastic wave of frequency v (or angular frequency $\omega = 2\pi v$) is

$$\varepsilon = (n + ½)\, hv = (n + ½)h\omega,$$

where $h = \frac{h}{2\pi}$ and n is the number of phonons in the mode, $h\omega$ is the quantum of elastic energy, that is, energy of a 'phonon' and ½ $h\omega$ is the zero-point energy of the mode.

The quantisation of lattice vibration energy has an important consequence when a lattice vibration is involved in an interaction. Since the energy of a mode of vibration has to be $(n+ ½)\, h_\omega$ with n any positive integer, and since to a first-order approximation the energy changes only to adjacent allowed values, we must have

$$\Delta\varepsilon = \pm\, h\omega$$

and correspondingly,

$$\Delta n = \pm 1.$$

This means that the energy change corresponds to a gain or loss of a single phonon at a time.

Experimental Evidence for Phonons : There is no direct experimental evidence that the energy of an elastic wave is quantised. (On the other hand, photoelectric effect is the direct experimental evidence that the energy of the electromagnetic wave is quantised). However, the following facts provide evidence for the existence of phonons :

(*i*) The lattice contribution to the heat capacity of solids approaches zero as the temperature approaches zero. This can be explained only if the lattice vibrations are quantised.

(ii) X-rays and neutrons are scattered inelastically by crystals. The change of energy and momentum in this process corresponds to the creation or absorption of one or more phonons. By measuring the energy and momentum of the scattered X-rays or neutrons we can determine the properties of phonons. Such experiments provide the strongest evidence for phonons.

Momentum of Phonon

A phonon of wave vector $\vec{K}$ interacts with other particles such as photons, neutrons and electrons as if it has a momentum $h\vec{K}$. This is in accordance with de Broglie relation $p = \frac{h}{\lambda} = \frac{h/2\pi}{\lambda/2\pi} = h\vec{K}$). Sometimes $h\vec{K}$ is called the 'crystal momentum'.

A phonon on a lattice, however, does not carry physical momentum, except for $\vec{K} = 0$ which corresponds to a uniform translation of the crystal as a whole. But for most practical purposes a phonon acts as if its momentum were *hK*.

In crystals there exist momentum conservation laws (wave vector selection rules) for allowed transitions between quantum states. The elastic scattering (Bragg diffraction) of an X-ray photon by a crystal is governed by the wave vector selection rule

$$\vec{k} = \vec{k} + \vec{G}$$

where $\vec{G}$ is a vector in the reciprocal lattice, $\vec{k}$ is the wave vector of the incident photon and $\vec{k}$ is the wave vector of the scattered photon.

If the scattering of the photon is inelastic, with the creation of a phonon of wave vector $\vec{k}$, then the wave vector conservation law (selection rule) becomes

$$\vec{k}' + \vec{K} = \vec{k} + \vec{G}.$$

If a phonon $\vec{k}$ is absorbed in the process, then the relation would be

$$\vec{k}' = \vec{k} + \vec{K} + \vec{G}.$$

Specific Heat of Crystalline Solids : The heat capacity of a solid at constant volume is defined as

$$Cv = \frac{DE}{dr},$$

where E is the internal energy and T the Kelvin temperature.

The experimental facts about the heat capacity of solids in general are the following.

(i) At ordinary and high temperatures the value of the heat capacity of nearly all solids is close to $3Nk$, where N is Avogadro's number and k is Boltzmann's constant. For 1 mole of atoms it is 6 cal/mole-K (Dulong and Petit law).

Dulong and Petit regarded each atom of the solid as executing simple harmonic vibrations about its lattice site in three dimensions. Thus one mole of the solid, having N atoms, has $3N$ degrees of freedom. According to the classical law of equipartition of energy, each degree of freedom is assigned an average total (kinetic + potential) energy kT. Therefore, the total energy of one mole of the solid is

$$E = 3NkT = 3\ RT$$

where R is the universal gas constant. Thus, the heat capacity at constant Volume is

$$C_v = \frac{dE}{dT} = 3R = 6 \text{ cal/mole-K.}$$

(ii) At lower temperatures the heat capacity drops markedly and approaches zero. Near the absolute zero, the heat capacity varies as T^3 in insulators and as T in metals. If the metal becomes a superconductor, the drop is faster than T.

(iii) In magnetic solids there is a large contribution to the heat capacity near the temperature at which the magnetic moments become ordered. Below 0.1 K the ordering of nuclear moments may give very large heat capacities.

The first and second points are illustrated in the figure below for silicon and germanium.

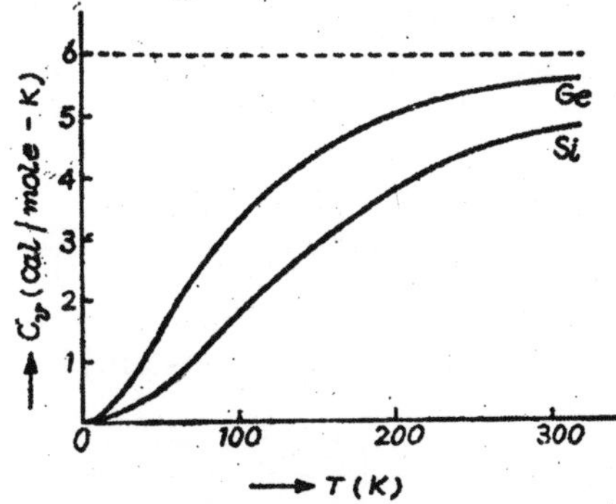

Quantum Theory of Einstein

Einstein, in 1907, applied quantum theory to the specific heat of solids. We know that the energy of a lattice vibration is quantised, the quanta of energy being 'phonons' which obey Bose-Einstein statistics. The energy of a phonon is $\hbar\omega$.

Einstein assumed that *all atoms of a solid lattice vibrate independently of each other with the same frequency in each of the three degrees of freedom.* Thus the energy of a lattice vibration depends only on its frequency ω (or phonon energy $\hbar\omega$) and phonon occupancy <*n*>, say. Now, one mole of the solid contains *N* atoms with 3*N* phonons of vibration, each with energy $h \hbar \omega$. Therefore, the energy per mole of the solid will be

$$E = 3N <n> \hbar \omega.$$

In thermal equilibrium at temperature *T* the occupancy is given by Bose-Einstein distribution

$$<n> = e^{(\epsilon - \epsilon F)/kT} \leq$$

and thus

$$E = 3N \hbar \omega \frac{1}{e^{\hbar\omega/kT} - 1}$$

The specific heat is given by

$$C_v = \frac{dE}{dT} = 3N \hbar \omega \frac{e^{\hbar\omega/kT}\left(\frac{\hbar\omega}{kT^2}\right)}{\left(e^{\hbar\omega/kT} - 1\right)^2}$$

$$= 3Nk \left(\frac{\hbar\omega}{kT}\right)^2 \frac{e^{\hbar\omega/kT}}{\left(e^{\hbar\omega/kT} - 1\right)^2}$$

But $Nk = R$ (the universal gas constant).

$$\therefore \quad C_v = 3R \left(\frac{\hbar\omega}{kT}\right)^2 \frac{e^{\hbar\omega/kT}}{\left(e^{\hbar\omega/kT} - 1\right)^2}.$$

Let us put $\frac{\hbar\omega}{k} = \Theta_E$, where Θ_E is known as Einstein characteristic temperature. Then we have

$$\boxed{C_v = 3R\left(\frac{\Theta E}{T}\right)^2 \frac{e^{\Theta E/T}}{\left(e^{\Theta E/T} - 1\right)^2}}$$

This is Einstein's specific heat formula.

(i) At high temperatures ($T >> \Theta_E$) we have $\frac{\Theta_E}{T} << 1$, so that $e\Theta E/T \approx 1 + \frac{\Theta_E}{T}$ and we have

$$C_v \approx 3R\left(\frac{\Theta_E}{T}\right)^2 \frac{1+\frac{\Theta_E}{T}}{\frac{\Theta_E}{T}} \approx 3R$$

$$= 6 \text{ cal/mole-K.}$$

This is Dulong and Petit law which agrees with experiment; at high temperatures.

(ii) At low temperatures ($T < < \Theta_E$) we have $\frac{\Theta_E}{T} >> 1$, so that $e^{\Theta_E/T} > > 1$, and the Einstein's formula gives

$$C_v \approx 3R\left(\frac{\Theta_E}{T}\right)^2 e^{-\Theta_E/T}$$ and we have

This shows that as $T \to 0$; the specific heat C_v also approaches zero, a conclusion in general agreement with experiment.

The figure below compares the experimental values of C, (shown by dots) for diamond with the behaviour predicted by Einstein's theory (solid curve), using the characteristic temperature $\Theta_E = h\omega/k = 1320$ K. The agreement is reasonably good. (Θ_E or ω is an adjustable parameter and will have different values for different materials to match the experimental results.).

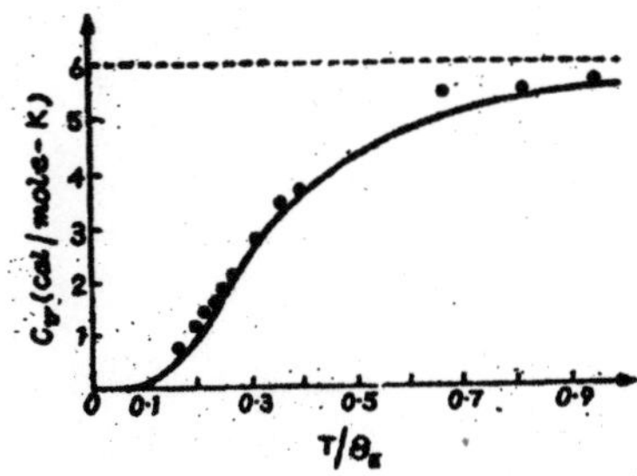

Limitation : There is no quantitative agreement at very low temperature. Near absolute zero, according to Einstein, $C_v \propto e^{-\Theta_E/T}$ whereas according to experiment, $C_v \propto T^3$. The reason is that Einstein oversimplified matters by assuming all of the phonons to have the same frequency.

In fact the phonon frequencies have a distribution like that in electromagnetic radiation in a cavity. This, in turn, is a result of the fact that the atomic vibrations in a lattice are strongly coupled and cannot be considered independent. The problem was taken up by Debye.

14

Linear Accelerator

Aim of Accelerators

Particle accelerators are machines for imparting high energies of the order of MeV, to more, to charged particles (electrons, protons, deuterons, α-particles), and also to bare nuclei of lighter atoms. Such energetic particles are required in making nuclear investigations, and in the production of transuranic elements.

There are three main types of particle accelerators : (*i*) electrostatic accelerators (such as Van de Graff generator); (*ii*) linear accelerators; and (*iii*) orbital accelerators (such as cyclotron, synchro-cyclotron, betatron, and the synchrotron).

Role of Linear Accelerator

A linear accelerator can impart energies to heavy ions, protons and electrons up to more than 10 MeV, which is the upper limit attainable by an electrostatic accelerator. In this accelerator the particle receives a series of additive accelerations in successive stages.

Construction : The principle diagram of an accelerator is shown in Fig. It consists of a long, evacuated tube carrying a series of coaxial cylindrical electrodes, 1, 2, 3, 4,..., of successively increasing length. Alternate sets of cylinders are connected together, and a high-frequency alternating p.d. is applied between the two sets by a radio-frequency oscillator. Thus alternate cylinders have opposite potentials. For example, in one half-cycle, cylinders, 1, 3, 5, ... are positive whilst 2, 4,,.. are negative; whereas in the next half-cycle 1, 3, 5,... become negative whilst 2,4, ...become positive.

Working : A charged particle keeps a constant speed while moving in the field-free space *within* a cylinder, but is accelerated while moving across the gap between two cylinders having a p.d.

Suppose that a positive particle enters cylinder 1 at a moment when the cylinders 1, 3, 5,... just acquire a negative potential, and the cylinders 2, 4,... a positive potential. The particle travels inside the cylinder with a constant speed. The frequency of the alternating p.d. applied between the two sets of cylinders is so chosen that the time taken by the particle to move through the cylinder is equal to the time of one half-cycle of the p.d. Thus as soon as the positive particle leaves the cylinder 1, the cylinders 1, 3, 5,... become positive and the cylinders 2, 4,... negative. The particle is therefore accelerated in the gap between cylinders 1 and 2, and enter the cylinder 2 with an increased speed. Once inside the cylinder 2, it travels with constant increased speed remaining unaffected by the changing potential of the cylinder. Since the successive cylinders are made longer, the particle takes the same time to traverse each cylinder in spite of its increased speed. Therefore, by the time the (positive) particle leaves the cylinder 2, the cylinders 2, 4,.... become again positive and the cylinders 1, 3, 5,... again negative. The particle is therefore again accelerated in the gap between 2 and 3 and reaches the cylinder 3 with a further increase in speed. The same happens at each successive gap right down the tube, and the particle goes on gaining energy. There is no theoretical limit to the final energy which is limited only by the length of the surrounding tube and the available power.

A typical length for a proton accelerator is between 5 and 30 meters. It can be fed with 4-5 MeV protons already accelerated by a Van de Graff generator, and can further increase their energy to about 50 MeV.

A linear accelerator can also impart energies to electrons. An electron, however, has a speed as high as 98% of the speed of light at an energy of only 2 MeV. The transit times through cylinders are therefore so short that extremely high frequency p.d.'s (≈ 1000 Mc/s) are required. Furthermore, beyond 2 MeV any further increase in electron speed is hardly possible, and any gain in energy comes rather from the relativistic *increase in mass* of the electron. The world's longest linear accelerator, about 3 km in length, delivers electrons with an energy of 30,000 MeV or more.

Significance of Cyclotron

A linear accelerator is inconvenient in shape. A more compact machine called a 'cyclotron' was devised by Lawrence and Livingston in 1932 to accelerate positive ions such as protons, deuterons and α-particles.

Constriction : It consists of two hollow D-shaped copper electrodes D_1 and D_2 (called the 'dees') with a small gap between them (Fig.). The dees are enclosed in an evacuated steel box placed between the poles of a very large magnet. The magnet produces a field B of about 15,000 gauss perpendicular to the plane of dees. A high-frequency alternating p.d. (≈ 12 Mc/sec $\approx 10^5$ volts) is applied across the dees by means of a powerful valve oscillator.

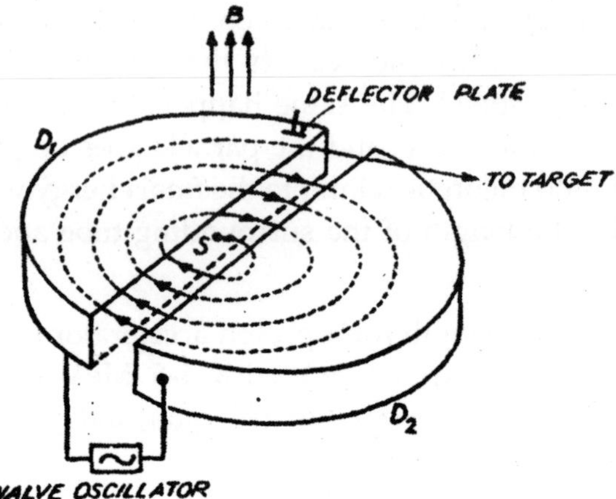

The ion source consists of a small chamber carrying a heated filament and the gas concerned (hydrogen for protons, heavy hydrogen for deuterons). The electrons emitted by the filament ionise the gas. The ions so produced pass through a capillary tube which ejects them at the centre 5 of the gap between the dees.

Working : Suppose that a positive ion is injected at S at a moment when D_2 just acquires a negative potential and D_1 a positive potential. The ion is then accelerated towards D_2 and enters it with increased velocity.

Once inside D_2 it becomes free from the electric field and travels with constant speed. Since the ion is moving in a magnetic field B perpendicular to the plane of dees, it adopts a circular path of radius r given by

$$\frac{mv^2}{r} = Bqv.$$

where m is the mass, q the charge and v the speed of the ion.

$$\therefore \qquad v = \frac{Bqr}{m}. \qquad \text{... }(i)$$

The time T required by the ion to complete one circular revolution is $2\pi r/v$. Therefore

$$T = \frac{2\pi r}{v} = \frac{2\pi m}{Bq}$$

The frequency of revolution is

$$f = \frac{1}{T} = \frac{Bq}{2\pi m}. \qquad (ii)$$

Thus, if B, m and q are constants, the period of revolution T, or the frequency, is independent of the radius of the path and of the speed of the ion. That is, all ions (same m, same q) 'take the same time to describe a circle (or semi-circle) irrespective of their speeds or energies. (In fact, it is this result which makes the cyclotron possible).

The frequency of the applied p.d. is adjusted to be equal to the frequency of revolution of the ions. When this is the case, the positive ion moving in D_2 completes a semi-circle and leaves it just at the moment when D_2 becomes positive and D_1 negative. The ion is therefore further accelerated in the gap and enters D_1. On account of its increased speed its semi-circular path in D_1 is of greater radius. *The time of passage through D_1, however, is still the same.* The process is repeated after every half-cycle of the p.d., and the ion gains speed each time it passes from one dee to the other. Finally, the ion reaches the outer edge of one dee where it is pulled out of the system by a negatively-charged deflector plate.

Evidently, eq. (*ii*), when f is taken as the frequency of the applied p d., is the resonance condition for the successful operation of the cyclotron.

The cyclotron has produced positive particles having energies of 20 MeV.

Energy Acquired

The velocity, and hence the kinetic energy, with which the

ion leaves the cyclotron can be obtained by putting $r = R$ (radius *of* dees) in eq (*i*). Thus

$$v = \frac{BqR}{m}$$

Therefore, the kinetic energy of the emerging ion is

$$K = \frac{1}{2}mv^2 = \frac{1}{2}m\left(\frac{BqR}{m}\right)^2$$

$$= \frac{1}{2}m\ B^2R^2\left(\frac{B}{m}\right)^2.$$

But $f = \dfrac{Bq}{2\pi\, m}$, where f is the frequency of the applied p.d.

Therefore $K = 2\pi^2 f^2\ m\ R^2$.

This is the expression for the energy of the particle in terms of the frequency of applied p.d. and the radius of the dees. The frequency, in turn, depends upon the flux density B. Thus in the non-relativist range the limitation on the maximum energy attained by a particular ion is determined by the size of the dees and the pole-faces of the magnet.

Limitation : The maximum energies of particles attainable in a cyclotron cannot be increased indefinitely by increasing the size of the apparatus because of the relativistic increase in mass of the particles in accordance with the equation

$$m = \frac{m_0}{\sqrt{\left(1 - \dfrac{v^2}{c^2}\right)}},$$

where m_0 is the rest mass of the particle and in the mass at speed v. As the speed v of the particle increases to become comparable with the speed of light c, its mass m increases appreciably. Therefore the frequency of revolution $f\left(\dfrac{Bq}{2\pi\, m}\right)$ goes on decreasing. Thus the particle traverses each dee too slowly and becomes more and more out of step with the applied p.d. until it can no longer be accelerated further.

At 20 MeV the mass of a proton is about 2% greater than its rest mass, but beyond this the increase in mass becomes serious enough to affect the operation of the cyclotron. Thus it is the upper limit of proton energy attainable by a cyclotron.

The electrons due to their very small mass approach the speed of light very quickly so that the relativistic increase in mass becomes appreciable at low enough energies. For example, the mass of an electron is doubled at an energy of only about 0.5 MeV. Hence the cyclotron *cannot* be used to accelerate electrons even to moderate energies.

Overcoming of Limitation (Synchro-cyclotron) : The energy limitation of 20 MeV resulting from the relativistic increase m the mass of the particle has been removed in a modified form of the cyclotron known as 'synchro-cyclotron' (or 'frequency-modulated cyclotron'). In this machine a high-frequency p.d. is applied at the moment of injection ; but as the frequency of revolution of the particle begins to slow down, the frequency of the p.d. is also reduced so that synchronism between the revolving particle and the accelerating p.d. is maintained. Using this technique, protons of 300 MeV energy and α-particles of 384 MeV energy have been obtained. The maximum energy is decided by the size of the magnet.

Betatron : The betatron is used to accelerate electrons to energies higher than the electrons (β-particles) emitted from radioactive substances. The betatron-accelerated electrons can produce X-rays so short in wavelength and so penetrating in character as to surpass the γ-rays from radioactive substances.

Principle : In the betatron the electrons are put in a *stable* circular orbit in which they are continuously accelerated due to an emf produced by an *increasing* magnetic flux through the orbit.

Let us consider an electron moving in a circular orbit of radius r in a magnetic field at right angles to the plane of the orbit (Fig.). The magnetic field is *non-uniform* in space, and *increasing* with time. Suppose, at a given instant, $\overline{B}$ is the average flux density *through* the orbit and B, the flux density *at the* orbit itself. The instantaneous magnetic flux Φ through the orbit will be given by

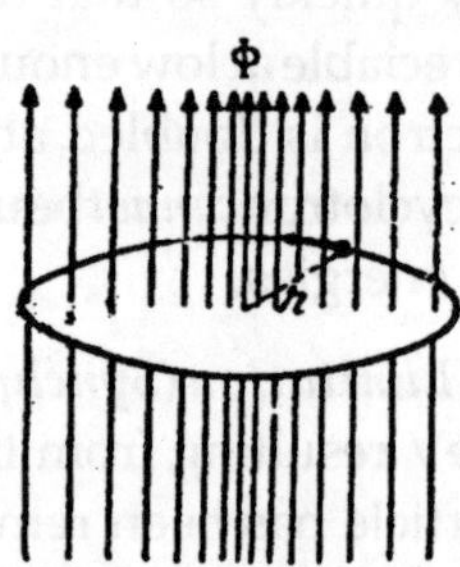

Φ = average flux density through the orbit × area enclosed by the orbit

$$= \overline{B} \times \pi r^2 = \pi r^2 \overline{B} \qquad \text{... } (i)$$

As the magnetic field is increasing with time, the magnetic flux Φ through the orbit is increasing at a rate $\frac{d\Phi}{dt}$. As a result, an emf is induced in the orbit. If V is the induced emf, then

$$V = \frac{d\Phi}{dt} \text{ (numerically).}$$

The work done on the electron in one complete revolution is

$$W = \text{charge on electron} \times \text{emf}$$

$$= eV.$$

Hence the tangential force acting on the electron is

$$F = \frac{\text{work done in one rev.}}{\text{distance moved in one rev.}}.$$

$$= \frac{W}{2\pi r} = \frac{eV}{2\pi r} = \frac{e}{2\pi r}\frac{d\Phi}{dt}.$$

Substituting the value of Φ from eq. (*i*), we obtain

$$F = \frac{e}{2\pi r}\frac{d\left(\pi r^2 \overline{B}\right)}{dt} = \frac{er}{2}\frac{d\overline{B}}{dt}. \quad \text{... (ii)}$$

The electron is accelerated by this force acting upon it. This force arises due to the emf produced by the *increasing* 'average flux density $\overline{B}$ through the entire orbital area. As $\overline{B}$ increases the electron energy increases and the electron tends to move it in an orbit of *increasing* radius.

Now, the electron is moving in a circular orbit of radius r in a magnetic field perpendicular to it. The instantaneous magnetic flux density at the orbit is B_r. If v be the instantaneous speed of the electron, the instantaneous radial force on it will be $B_r ev$. This must be equal to the centripetal force mv^2/r. Thus

$$Brev = \frac{mv^2}{r}.$$

Therefore, the instantaneous momentum of the electron is

$$p = mv = B_r er.$$

The rate of change of momentum is

$$\frac{dp}{dt} = \frac{d\,(b_1 er)}{dt} = er\,\frac{dBr}{dt}.$$

By Newton's law this must be equal to a tangential force F acting on the electron. Thus

$$F = er\frac{dBr}{dt}. \quad \text{... (iii)}$$

Clearly this force arises by the *increasing* flux density B_r at the orbit. As B_r increases, the electron tends to move in an orbit of *decreasing* radius ($B_r\ ev = mv^2/r$).

Thus the forces given by eq. (*ii*) and (*iii*) have opposite actions on the radius of the orbit ; the former tends to increase while the latter tends to decrease the radius Hence, to preserve a constant radius of the orbit these two forces must be equal numerically. That is

$$\frac{er}{2}\frac{d\overline{B}}{dt} = er\frac{dB_r}{dt}.$$

Integrating both sides of the equation, assuming that $\overline{B} = B_r = 0$ at $t = 0$, we obtain

$$\frac{er\overline{B}}{2} = erB_r$$

or $$\overline{B} = 2B_r. \qquad \text{(Betatron condition)}$$

Thus the average magnetic flux density through the orbit must always be twice the flux density at the orbit. This is the condition of betatron. This condition is achieved by suitably designing the size and shape of the magnet providing the magnetic field. When this is done, the electrons once put in the orbit go again and again around the same orbit with continuous increasing speed so long as the magnetic field is increasing (with time).

In betatron the increase in speed around the circular orbit is *continuous* and there is no synchronism to be maintained. Hence any relativistic increase in mass of the electron and the consequent slowing down of the rate of revolution do not affect the operation of the betatron.

Construction : The construction of the betatron is illustrated in Fig. It consists of a" large highly-evacuated annular tube *DD*, called a "donut" tube, in which the electrons revolve. Their orbit is a circle at right angles to the plane of the page. They are shown entering the page at the right (e) and emerging the page at the left (x). The tube *DD* is placed between the specially-designed pole-pieces of a large electromagnet energised by an alternating current. The magnet satisfies the betatron condition by providing a larger flux density in the central part of the field. Since the magnet is energised by *alternating* current, an increasing magnetic field in a given direction is obtained only during one-quarter cycle in which the current increases from zero to the peak value. Hence the electrons are kept in their orbit only during this part of each cycle, and the output electron beam is in pulses.

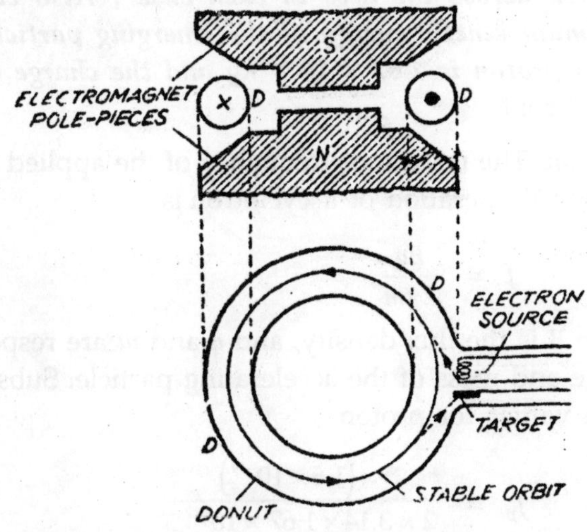

Action : The plane of the orbit is shown in the diagram. The electrons to be accelerated are injected into the donut tube *D* by means of an electron-gun. The gun is so *pulsed* that the electrons are injected only at the start of a quarter-cycle in which the magnetic field increases in a given direction from zero to the peak value. These electrons are soon captured into a *stable* orbit and make several hundred thousand revolutions during the quarter-cycle, being accelerated continuously. At the end of each accelerating quarter-cycle the orbit is momentarily expanded by means of an auxiliary magnetic field (produced by sending a pulse of an extra current through an auxiliary coil), so that the electrons strike the target or emerge through a window.

Betatrons have produced electrons with energies of greater than 300 MeV.

PROBLEMS

1. A cyclotron in which the magnetic flax density is 3 weber/ m^2 and whose dees' radius is 0.5 m, is used to accelerate (i) protons, (ii) α-particles. What must be the frequency of the p.d. applied across the dees in each case ? Also calculate the maximum kinetic energy of the emerging particles. The mass of a proton is 1.67×10^{-27} kg and the charge on it is 1.6×10^{-19} coul.

Solution: The resonance frequency of the applied p.d. for the successful operation of a cyclotron is

$$f = \frac{Bq}{2\pi m}, \qquad \text{... (i)}$$

where B is the flux density, and q and m are respectively the charge and mass of the accelerating particle. Substituting the given values for proton

$$f_p = \frac{3 \times (1.6 \times 10^{-19})}{2 \times 3.14 \times 1.67 \times 10^{-17}}$$

$$= 46 \times 10^6 \text{ cycles/sec}$$

$$= 46 \text{ mc/sec.}$$

The mass of an α-particle is 4 times that of proton, while the charge is twice that of proton. Hence, by eq. (i), the required frequency is

$$fa = \frac{1}{2} f_p = 23 \text{ mc/sec.}$$

To find the maximum energy of the emerging particles let us consider the relation

$$\frac{mv^2}{r} = Bqv$$

$$\text{so that} \quad v = \frac{Bqr}{m}$$

If $r = R$, the radius of dee, then v would represent the (maximum) velocity of the emerging particle. Thus, the maximum kinetic energy is

$$K = \frac{1}{2} mv^2 = \frac{1}{2} m \left(\frac{Bqr}{m}\right)^2$$

$$= \frac{1}{2} \frac{B^2 q^2 R^2}{m}. \qquad \text{... } (ii)$$

Substituting the given values for protons,

$$K_p = \frac{1}{2} \times \frac{(3)^2 \times \left(1.6 \times 10^{-19}\right)^2 \times (0.5)^2}{1.67 \times 10^{-27}}$$

$$= 1.72 \times 10^{-11} \text{ joule.}$$

$$\text{Now } 1 \text{ eV} = 1.6 \times 10^{-19} \text{ joule.}$$

$$\therefore \quad K_p = \frac{1.72 \times 10^{-11}}{1.6 \times 10^{-19}} = 1.1 \times 10^8 \text{ eV} = 110 \text{ MeV.}$$

It can be seen from eq. (*ii*) that the kinetic energy of the emerging α-particles (charge double, mass four times than proton) would be the same as for protons. Thus

$$K_\alpha = K_p = 110 \text{ MeV.}$$

2. A cyclotron is being used for accelerating α-particles. The frequency of potential on the dees for α-particles has been 16 mc/sec and maximum energies of 48 MeV have been attained. What change in frequency, if any, is needed to accelerate deuterons and what maximum energies may be attained ? (The mass, as well as the charge, of a deuteron is half of that of an α-particle).

[**Ans.** No change in frequency, 24 MeV]

3. The diameter of a cyclotron is 18 m and the strength of magnetic field is 0.7 weber/m^2. Calculate the energies to which (i) a proton, (ii) a deuteron can be accelerated. Mass of a proton, m_p= 1.008 amu ; charge of an electron, e=1.6 × 10^{-19} coulomb and 1 amu = 1.66 × 10^{-27}kg.

Solution: When the particle describes a circle of radius *R*, equal to Dees' radius, the velocity reaches maximum value. Equating the magnetic force to the centripetal force, we have

$$Bqv = \frac{mv^2}{R}$$

or
$$v = \frac{BqR}{m}.$$

∴ the (maximum) kinetic energy is

$$K = \frac{1}{2}mv^2$$

$$= \frac{B^2q^2R^2}{2m} \quad \text{... } (i)$$

The charge on a proton is + 1.6 × 10^{-19} coul (equal to the negative charge of electron) and its mass is 1.008 × (1.66 × 10^{-27} kg) = 1.67 × 10^{-27} kg. Therefore, the maximum kinetic energy of the proton is

$$K_p = \frac{(0.7\ \text{weber}/\text{m}^2)^2 (1.6\times 10^{-19}\ \text{coul})^2 (0.9\ \text{m})^2}{2\times 1.67\times 10^{-27}\ \text{kg}}$$

$$= 3.0 \times 10^{-12}\ \text{joule}$$

$$= \frac{3.0\times 10^{-12}}{1.6\times 10^{-19}} = 1.9 \times 10^7\ \text{eV} = 19\ \text{MeV}.$$

The charge on deuteron is same as on proton, but its mass is almost double than that of proton. Hence, as seen from eq. (*i*) its kinetic energy will be almost half than that of proton.

$$\therefore \quad K_\alpha = 9.5\ \text{MeV}.$$

4. A cyclotron with a radius of 10 m has a magnetic field of 2.0 T (tesla). Determine the maximum energy of the accelerated deuterons. Through what potential difference would they have to travel to attain the energy ? The mass of deuteron is 3.34 x 10^{-27} kg and electronic charge is 1.6 × 10^{-19} coulomb.

Solution: The maximum kinetic energy of the deuterons is given by

$$K = \frac{B^2 q^2 R^2}{2m}$$

Substituting the given values, remembering that the positive charge on deuteron is equal to the negative charge on electron, we get

$$K = \frac{(2.0 \text{ tesla})^2 (1.6 \times 10^{-19} \text{ coulomb})^2 (1.0 \text{ m})^2}{2 \times 3.34 \times 10^{-27} \text{ kg}}$$

$$= 1.53 \times 10^{-11} \text{ joule}$$

$$= \frac{1.53 \times 10^{-11}}{1.6 \times 10^{-19}} = 0.96 \times 10^8 \text{ eV} = 96 \text{ MeV}.$$

To achieve this energy, the deuteron will have to travel through a p.d. of 96 million volts.

5. Calculate the maximum energy which a cyclotron can give to α-particles if the diameter of dee is 1.2 meter and the magnetic field is 1′5 wb/m^2. At what frequency the cyclotron be operated to achieve this energy ? The mass of an α-particle is 6.645 × 10^{-27} kg and the charge on it is 3.2 × 10^{-19} coulomb. (1 MeV = 1.6 × 10^{-13} joule).

[**Ans.** 39 MeV, 11.5 Mc/sec.]

6. A cyclotron whose magnetic pole faces have a diameter of 1.52 meters is operated with a radio-frequency field at 10^7 cycles per sec. Show that it can accelerate protons to an energy of approximately 12 MeV and α-particles to approximately 48 MeV. The mass of a proton is 1.67 × 10^{-27}

kg and that of an α-particle roughly four times.[Hint : K= 2 $\pi^2 m f^2 R^2$

7. A cyclotron of diameter 1 meter is used to accelerate protons. Across it is applied an alternating voltage of 10 mc/ sec frequency and of peak value 5000 volts. Calculate the magnetic field needed and the energy of the emergent beam. The mass and charge of proton are 1.67×10^{-27} kg and 1.6 × 10^{-19} coulomb respectively.

Solution: The resonance frequency of the applied voltage for the successful operation of the cyclotron is

$$f = \frac{Bq}{2\pi m},$$

so that the required field is

$$B = \frac{2\pi mf}{q}$$

Substituting the given values :

$$B = \frac{2 \times 3.14 \times \left(1.67 \times 10^{-27}\ \text{kg}\right) \times \left(10 \times 10^{6}\ \text{sec}^{-1}\right)}{1.6 \times 10^{-19}\ \text{coul}}$$

$$= 0.66\ \text{weber/meter}^2.$$

The kinetic energy of the emerging proton is

$$K = \frac{B^2 q^2 R^2}{2m}$$

$$= \frac{\left(0.66\ \text{weber/m}^2\right)^2 \times \left(1.6 \times 10^{-19}\ \text{coul}\right)^2 \times \left(0.5\ \text{meter}\right)^2}{2 \times \left(1.67 \times 10^{-27}\ \text{kg}\right)}$$

$$= 8.0 \times 10^{-13}\ \text{joule} = \frac{8.0 \times 10^{-13}}{1.6 \times 10^{-13}} = 5\ \text{MeV}.$$

8. Calculate the energy of a proton when it emerges out of a cyclotron after 50 revolutions. The voltage applied across the dees is 20 kV.

Solution: The proton (charge *e)* passes across the gap between the two dees twice in each revolution, and acquires energy *eV* during each transit, where *V* is the peak value of the voltage applied. Thus the energy acquired in 50 revolutions is

$$K = 2\ \text{eV} \times 50 = 100\ eV$$

$$= 100 \times (1.6 \times 10^{-19}\text{coul}) \times 20{,}000 \text{ volts}$$

$$= 3.2 \times 10^{-13} \text{ joule.}$$

$$\text{Now, } 1 \text{ MeV} = 1.6 \times 10^{-13} \text{ joule.}$$

$$\therefore \quad K = \frac{3.2 \times 10^{-13}}{1.6 \times 10^{-13}} = 2 \text{ MeV.}$$

9. ***If the maximum r. f. voltage applied to a cyclotron be 10 kV, show that a proton has to make 900 revolutions within the cyclotron before achieving $\frac{1}{5}$th the velocity of light. Take proton mass 1.6 × 10^{-27} kg and charge 1.6 × 10^{-19} coulomb. The velocity of light is 3 x 10^8 m/sec.***

10. ***An alternating potential difference of 50 kV maximum value is applied to the dees, whose radius is 40 cm, of a cyclotron by an oscillator. A deuteron of mass 2 amu acquires energy of 4 MeV in the cyclotron. Calculate (i) magnetic field strength, (ii) frequency of oscillator, (iii) number of revolutions which the deuteron has to make inside the cyclotron to gain energy. (1 amu= 1.6 × 10^{-27} kg, e = 1.6 × 10^{-19} coulomb).***

Solution: The expression for the kinetic energy of particles of mass *m* and charge *q* emerging from a cyclotron is

$$K = \frac{B^2 q^2 R^2}{2\,\text{m}},$$

where *B* is the magnetic field and *R* the dees radius.

Here $q = e = 1.6 \times 10^{-19}$ coul,

$$R = 0.4 \text{ meter,}$$

$$m = 2\text{amu} = 2 \times 16 \times 10^{-27} \text{ kg},$$

and $$K = 4\text{MeV} = 4 \times 1.6 \times 10^{-13} \text{joule}.$$

Thus

$$B^2 = \frac{2\,Km}{q^2 R^2}$$

$$= \frac{2 \times \left(4 \times 1.6 \times 10^{-13}\right) \times \left(2 \times 1.6 \times 10^{-27}\right)}{\left(1.6 \times 10^{-19}\right)^2 \times (0.4)^2} = 1.0.$$

$$\therefore \quad B = 1.0 \text{ weber/m}^2.$$

The (resonance) frequency of p. d. required for the successful operation of the cyclotron is

$$f = \frac{Bq}{2\,\pi\,m} = \frac{1.0 \times \left(1.6 \times 10^{-19}\right)}{2 \times 3.14 \times \left(2 \times 1.6 \times 10^{-27}\right)}$$

$$= 8 \times 10^6 \text{ cycles/sec} = 8 \text{ mc/sec}.$$

Suppose the deuteron makes n revolutions to gain energy K. In each revolution it is pushed twice and in each push gains energy qV, where V is the peak value of the applied alternating voltage. Thus

$$n \times 2qV = K$$

$$\therefore \quad n = \frac{K}{2\,qV}$$

$$= \frac{4 \times 1.6 \times 10^{-13} \text{ joule}}{2 \times \left(1.6 \times 10^{-19} \text{ coul}\right) \times (50{,}000 \text{ volts})} = 40.$$

11. In a certain betatron the maximum magnetic field at the orbit was 0-4 tesla, operating at 50 cps with a stable orbit diameter of 1.5 meter. Calculate the average energy gained per revolution and the final energy of the electrons. Assume that the electrons move with nearly the velocity of light.

Solution: In betatron the electrons move with velocity nearly c so that the total distance travelled in the acceleration time (one-quarter cycle) is

$$d = c \times \frac{T}{4} = \frac{c}{4f},$$

where f is the frequency of the A.C. supply feeding the magnet. If r be the radius of the stable orbit, then the total number of revolutions made by the electrons in the acceleration time is

$$n = \frac{d}{2\pi r} = \frac{c}{8\pi f r}.$$

Substituting the given values :

$$n = \frac{3 \times 10^8 \text{ m/sec}}{8 \times 3.14 \times 50/\text{sec} \times 0.75 \text{ m}} = 3.17 \times 10^5.$$

Relativistically, the momentum of the electron is given by

$$p = \frac{E}{c}$$

where E is the final energy acquired. If B is the maximum magnetic field *at the orbit*, and v is the final velocity acquired, then we have

$$Bev = mv^2/r,$$

where m is the (relativist) mass of the electron at velocity v. Thus

$$mv = Ber$$

or $$p = mv = Ber = \frac{E}{c}$$

or $$E = Berc.$$

Substituting the given values, the final energy acquire is

$$E = (0.4 \text{ weber/m}^2) \times (1.6 \times 10^{-19} \text{ coul}) \times (0.75\ m) \times (3 \times 10^8 \text{ m/sec})$$

$$= 1.44 \times 10^{-11} \text{ joule}$$

$$= \frac{1.44 \times 10^{-10}}{1.6 \times 10^{-13}}$$

$$= 90 \text{ MeV.} \quad [\therefore 1 \text{ Mev} = 1.6 \times 10^{-13} \text{ joule}]$$

The average energy per revolution is

$$\frac{90 \times 10^6 \text{ eV}}{3.17 \times 10^5} = 284 \text{ eV.}$$

12. Calculate average energy per revolution and final energy gained by electrons in a betatron to which is applied a maximum magnetic field of 5000 gauss operating at 60 hertz in a stable orbit of diameter 2 meters. (1 gauss=10^{-4} tesla, $c = 3 \times 10^8$ m/sec, $e = 1.6 \times 10^{-19}$ coul and 1 MeV = 1.6×10^{-13} joule).

[Ans. 750 eV, 750 MeV]

Bibliography

Alam, T.: *Physical Principles of Flow in Unsaturated Porous Media,* Oxford, New York, 1994.

Arun, Kumar: *Problems in Physics,* Atlantic Publishers, New Delhi, 2003.

Atherton, J.C.: *Thermal Physics,* Butterworths, London, 1987.

Bowling, D.J.F.: *Basic Physics,* Chapmann and Hall, London, 1976.

Caro, D. E.: *Modern Physics: an Introduction to Atomic and Nuclear Physics,* Edward Arnold, London, 1985.

Cartwright, H.R.: *Elements of Thermal and Statistical Physics,* Oxford University Press, London, 1966.

Crafts, A.S.: *Phloem Transport in Plants,* W.H. Freeman, San Francisco, 1973.

Daniel, W.: *Introduction to Physics,* MacMillan Press, Melbourne, 1969.

Divan, T.: *Thermal Analysis of Materials,* W.A. Benjamin, California, 1975.

Donald, J.S.: *Advanced Guide to Modern Physics,* Pelham Books, London, 1976.

Epstein, E.: *Statistical Physics,* John Wiley and Sons, New York, 1972.

Esau, K.: *Physical Geology,* John Wiley and Sons, New York, 1965.

Feynman and Others: *The Feymman Lectures on Physics,* Add. Wesley, London, 1989.

Feynman, R. P. and Others: *Elementary Particles and the Laws of Physics,* Cambridge, New York, 2002.

Francia, G.: *The Investigation of the Physical World,* Cambridge, New York, 1981.

Ftzler M.E.: *Plant Lectins: Molecular and Biological Aspects,* Academic Press, New York, 1985.

George, M.: *Elements of Physical Hydrology,* Johns Hopkins, New York, 1998.

Green, J.R.: *Modern Physics,* MacMillan, London, 1911.

Halevy, A.H.: *Handbook of Physics,* C.R.C. Press, Boca Raton, 1985.

Hart, J.W. and Sabnis, D.D.: *Encyclopaedia of Stastical Physics,* Springer Verlag, Berlin, 1982.

Hopkins, C.R.: *Structure and Function of Cells,* W.B. Saunders Co., London, 1978.

Jorgensen, T. P.: *The Physics of Golf,* Springer, New York, 1999.

Karl, F.: *Basic Physics,* John Wiley, London, 1996.

Kinet, J.M.: *Thermal and Statistical Physics,* C.R.C. Press, Boca Raton, 1981.

Kirkby, E.A.: *Mathematical Physics,* The University of Leeds, Leeds, 1970.

Lain, G.: *Vibrations and Waves in Physics,* Cambridge, New York, 1998.

Landshoff, P.: *Essential Quantum Physics,* Cambridge, New York, 2001.

Lipson, S. G.: *Optical Physics,* Cambridge, New York, 1995.

Lockett, Keith: *Physics in the Real World,* Cambridge, New York, 1995.

Lui, Lam: *Introduction to Nonlinear Physics,* Springer, New York, 1997.

McKee, H.S.: *University Physics,* Oxford University Press, London, 1962.

Meidner, H. and Sheriff, D.W.: *Water and Plants,* Blackie and Sons, Glasgow, 1976.

Mengel, K. and Kirkby, E.A.: *Thermal and Statistical Physics,* Praeger Publishers, New York, 1974.

Mercer, E.I.: *Recent Advances in the Chemistry and Biochemistry of Plant Lipids,* Academic Press, New York, 1975.

Minoru, Fujimoto: *The Physics of Structural Phase Transitions,* Springer, London, 2000.

Nabapro, F.R.: *The Physics of Creep,* T. & F. Publications, New York, 1995.

Negele, J. W. and Others: *Advances in Nuclear Physics,* Plenum, London, 1994.

Palladin, V.I.: *Physical Metallurgy,* Arihant Publishers, Jaipur, 1988.

Peter, Haasen: *Optical Physics,* Cambridge, New York, 1996.

Plischke, M. and Others: *Equilibrium Statistical Physics,* World Scientific, New York, 1994.

Pruce, D.: *Elements of Stastical and Thermodynamics Physics,* McGraw Hill, London, 1975.

Rabinowitch, E.: *Thermodynamics and Statistical Physics,* John Wiley, New York, 1969.

Rawn, J.D.: *Encyclopaedia of Modern Physics,* Harper and Row, New York, 1983.

Ross, C.W. and Salisbury, F.B.: *Thermal Physics,* Woodworth Publishing Co., California, 1967.

Rossi, Paolo: *The Birth of Modern Science,* Blackwell, New Jersey, 2001.

Ruhland, W.: *Encyclopaedia of Thermodynamics Physics,* Springer-Verlag, Berlin, 1961.

Sagdeev, R.Z.: *Nonlinear Space Plasma Physics,* AIP, New Jersey, 1993.

Schein, L. B.: *Electro-Photography and Development Physics*, Springer, London, 1992.

Slatyer, R.O.: *The Physics of Structural Phase Transitions*, Academic Press, New York, 1967.

Srivastava, H.S.: *Elements of Physics*, Rastogi Publications, Meerut, 1990.

Thimann, K.V.: *Physics and the Real World*, CRC Press, Boca Raton, Florida, 1980.

Tolbert, N.E.: *Microbodies—Peroxisomes and Glyoxysomes*, Oxford University Press, London, 1971.

Tommy, L.: *Modern Physics in Twenty first Century*, Springer Verlag, Berlin, 1975.

Tucker, G.A.: *Physics and Chance*; Oxford, New York, 1999.

Verma, W.A.: *Problems in Physics*, Academic Press, New York, 1970.

Verner, J.E.: *Thermal Physics*, Academic Press, New York, 1985.

William, R.: *Techniques for Nuclear and Particle Physics Experiments: A How to Approach*, Springer, London, 1994.

Young, H. D.: *University Physics*, Add. Wesley, London, 1996.

Zeevart, J.A.: *Thermal Plasma and New Materials Technology*, McGraw Hill, London, 1976.

Zenic, B.: *The Geometry and Physics of Knots*, Cambridge, New York, 1991.

Index

E

F

G

H

I

J

K

L

M

N

O

P

Q

R

S

T

U

V

W

X

Y

Z

❑❑❑